T0211655

Managing Flood Risk

Anna Serra-Llobet • G. Mathias Kondolf
Kathleen Schaefer • Scott Nicholson
Editors

Managing Flood Risk

Innovative Approaches from Big Floodplain
Rivers and Urban Streams

palgrave
macmillan

Editors
Anna Serra-Llobet
University of California Berkeley
Berkeley, CA, USA

G. Mathias Kondolf
University of California Berkeley
Berkeley, CA, USA

Institut d´Études Avancées
Aix-Marseille Université
Marseille, France

Institut d´Études Avancées
Université de Lyon
Lyon, France

Kathleen Schaefer
University of California Berkeley
Berkeley, CA, USA

Scott Nicholson
University of California Berkeley
Berkeley, CA, USA

Federal Emergency Management
Agency, Oakland, CA, USA

US Army Corps of Engineers
Washington, DC, USA

ISBN 978-3-030-10092-6   ISBN 978-3-319-71673-2   (eBook)
https://doi.org/10.1007/978-3-319-71673-2

Cover illustration: Mono Circles © John Rawsterne/patternhead.com

Printed on acid-free paper

This Palgrave Macmillan imprint is published by the registered company Springer International Publishing AG part of Springer Nature
The registered company address is: Gewerbestrasse 11, 6330 Cham, Switzerland

FOREWORD

Rivers and their floodplains are integral features of our landscape. They create and tie together our topography, they provide conduits for transporting water and sediment downstream, and they complete the integration of the terrestrial and aquatic worlds. Historically seasonal floods would replenish and rework soils and nutrients, provide critical water for the riparian vegetation, and continually reshape and define the geomorphic character of the river. As human progress expanded, rivers and floodplains became the focal point for development. With that development came the desire and need to control the rivers and floods, often resulting in the construction of levees, dams, and river control features. These engineering developments resulted in floodplains worldwide being physically separated from the rivers that shaped and maintained them—a disjointed system.

Today new challenges are facing communities and economies that depend on the rivers and floodplains. For the last 200 years we have focused on trying to control floods with varying levels of success. What was forgotten is that local engineering structures alone will not provide the long-term protection as hydrology and river dynamics change in response to increased climate variability. Today the costs of floods are increasing exponentially as more populated areas are inundated, structures are lost, and societal infrastructure is impacted. Traditional engineering approaches to managing, adapting, and mitigating flood events, which have largely relied on flood control structures and insurance for recovery, are no longer providing the protection or reducing the risk of living in a historical floodplain.

This book summarizes experiences from large and small rivers and urban and developed floodplains in the USA and Europe. A couple of lessons immediately jump out from the case studies. One is that it is imperative that we do not continue to rely only on engineering fixes to protect people and development. We have to understand that the challenges facing water managers now are much larger and complex than traditional engineering approaches. Secondly, it is imperative that reducing the risk and resultant costs of flooding requires that people think about river basin management rather than site-specific management. Understanding how a watershed and river work together will help to prioritize specific floodplain management activities and reduce the overall risk and cost to people and local economies. Lastly, floodplain and river management is not just the province of city and state water managers. Today a comprehensive approach is needed, one that includes science, engineering, risk management, infrastructure protection, and land-use planning, along with an educational program for locally impacted residents.

Recent severe flooding in Texas, Florida, and elsewhere, coupled with rising sea levels and more intense storms, only increases the urgency for better flood risk management and improved floodplain land use in particular. It is essential that we take next steps to integrate what we have learned so far from innovative approaches around the globe and convert this into better policy and better planning.

Jacobs Engineering   David L. Wegner
Formerly Professional Staff,
House Transportation & Infrastructure
Subcommittee on Water Resources & Environment
Dallas, TX, USA

ACKNOWLEDGMENTS

We are indebted to the Institute of International Studies at the University of California, Berkeley (UC Berkeley) for its support of our research on flood risk management and its evolution in the US and Europe. Our comparative studies have been informed by the opportunity to host scholars from Washington, DC; various other points across the US; Brussels; and other European capitals, notably in workshops and seminars over the period 2012–2016 held as part of the interdisciplinary faculty seminar *Water Management: Past and Future* held at the Institute. Comparative study of flood risk management approaches now being implemented under the Floods Directive in the European Union with those still dominant in the US can provide a mirror through which we can better understand current problems in the US, as reflected in the Mississippi River Basin, the Sacramento Valley, and elsewhere.

Serra-Llobet is indebted to the Institut d'Études Avancées d'Aix-Marseille Université and the Laboratoire d'Excellence (Labex) OT-Med in Aix-en-Provence and Kondolf is indebted to the Collegium – Institut d'Études Avancées de Lyon, the European Institutes for Advanced Study (EURIAS) Fellowship Programme, and the European Commission's Marie Skłodowska-Curie actions (COFUND Programme – FP7), respectively, for the opportunity to complete this book in environments highly supportive of research and a wonderful quality of life.

We thank Mark R. Tompkins (FlowWest, Oakland) and Shana Udvardy (Union of Concerned Scientists) for their contributions to a UC Berkeley workshop on flood risk management involving most of the contributors to this book. We thank Jason Alexander for tracking down a high-resolution image of the photo of Mississippi River Lock and Dam 27 (Fig. 2.6). Laurent Schmitt thanks Pr. Jean-Nicolas Beisel, Pr. Karl M. Wantzen, Dr. Ulrike Pfarr, and Mr. Marc Lebeau for fruitful scientific and technical discussions on these topics.

We thank our contributors for their participation in our discussions and their written contributions. Of our 20 contributors, 17 are primarily from practice, working as designers, planners, engineers, public agency staff, consultants, or with NGOs. We are grateful for their efforts to convey experiences from the 'front line' of implementing flood risk management in this time of rapidly changing practice, a nice balance to the more academic perspective that (understandably) dominates most of the literature on the topic. Thanks to the willingness of our contributors to share their experiences, this book provides a unique perspective on challenges—and also opportunities—to implementing true flood risk management within the North American and European context.

CONTENTS

BIOGRAPHIES

EDITORS

Anna Serra-Llobet Director Sustainable Floodplains Initiative, Institute of International Studies, University of California Berkeley, and Research Fellow at the Institute for Advanced Studies, Aix-Marseille University, France. Serra-Llobet is an environmental scientist specializing in flood risk management (in the EU, USA, and elsewhere) and green infrastructure implementation in Latin America. Her dissertation analyzed the effect of the Tous Dam failure on flood management policy in Spain and the evolution of flood risk, vulnerability, and resilience in communities along the coast of NE Spain. She worked in the Water Unit of the Directorate-General for Environment of the European Commission (2011), was a Prometeo visiting professor at University of Cuenca, Ecuador (2014), and is conducting a comparative analysis of flood risk management policies in the EU and USA.

G. Mathias (Matt) Kondolf Professor Environmental Planning, Dept. Landscape Architecture & Environmental Planning, University of California Berkeley, and EURIAS Research Fellow at the Collegium—Institute for Advanced Studies at the University of Lyon, France. Kondolf is a fluvial geomorphologist specializing in environmental river management and restoration. At Berkeley he teaches courses in hydrology, river restoration, and environmental science. His research focuses on human-river interactions, with an emphasis on managing flood-prone lands, managing sediment in rivers and reservoirs, and river restoration. He is a co-editor of *Tools in Fluvial Geomorphology*. He has served as an adviser to

USA and state agencies on river management and restoration, and provided expert testimony before the US Congress, the California Legislature, and the International Court of Justice and Permanent Court of Arbitration in The Hague.

Co-editors

Kathleen Schaefer Researcher, Center for Catastrophic Risk Management, University of California Berkeley, Federal Emergency Management Agency (FEMA, retired). Schaefer is a researcher specializing in flood risk mapping and assessments with the UC Berkeley Center for Catastrophic Risk Management. She is the chair of the San Francisco Bay Coastal Hazards Adaption Resiliency Group (CHARG), an affiliation of federal, state, and local agencies working to develop a suite of climate change adaptation strategies for the San Francisco Bay Shoreline. For almost a decade, she worked for FEMA overseeing the production of Flood Insurance Rate Maps for Northern California.

Scott Nicholson Researcher, Center for Catastrophic Risk Management, University of California Berkeley, and Senior Policy Advisor and Analyst, Civil Works Policy and Planning Division, US Army Corps of Engineers, Washington, DC. Nicholson conducts research on flood risk management in the context of national policy, and advises on policy and on the development of Policy and Planning initiatives associated with navigation, flood risk management, and ecosystem restoration business lines. Nicholson is a graduate of the University of California Berkeley with an MS in Civil Engineering; Masters of City Planning (MCP) and Masters of Landscape Architecture (MLA). He formerly served as a staff member on the House Transportation and Infrastructure Committee, contributing to the development of Water Resources Development Acts. He has led complex state and federal environmental programs and projects through intergovernmental planning and construction efforts and formerly served as a civil works planning manager for the Pacific Ocean Division Regional Integration Team.

Contributors

Mitch Avalon Contra Costa County Flood Control & Water Conservation District (retired). Mitch Avalon is a graduate of the University of California, Berkeley where he received his Bachelor of Science Degree in Civil

Engineering. He joined Contra Costa County Public Works Department in 1979. Mr. Avalon has worked in many areas of Public Works, including transportation engineering, development engineering, design, and construction. Mr. Avalon recently retired as Deputy Public Works Director and Deputy Chief Engineer for the Flood Control and Water Conservation District, where he oversaw the development of regional management plans for flood protection and enhancement of the environmental resources in the District's 120 km of creeks and 30 flood management basins, and continues to advise the District in these matters. Since 1997, Mr. Avalon has chaired the Alhambra Watershed Council, whose 2001 Watershed Management Plan was based on a community based planning process and is now being implemented.

John Cain American Rivers, MLA, is the director of Conservation for California Floodplain Management for American Rivers. Over the last 20 years, Cain's work has focused on the interrelated issues of river restoration, water supply management, and flood risk reduction in California's Central Valley and Delta, where he has advanced several large-scale habitat restoration efforts including the 1200-acre Dutch Slough tidal marsh restoration project and rewatering of the San Joaquin River. In 2017 he was honored as Floodplain Manager of the Year by the Flood Management Association, the California Chapter of the Association of State Floodplain Managers, for his work to integrate ecosystem restoration into the Central Valley Flood Protection Plan. He previously served as the restoration ecologist for the Natural Heritage Institute and staff scientist for the Mono Lake Committee. He holds a bachelor's degree in physical geography, an MLA, and a master's degree in environmental planning from UC Berkeley.

Jack Curley Marin County Flood Control and Water Conservation District (retired). Curley was Capital Planning and Project Manager for the Ross Valley Watershed Program, Marin County, California. After spending 10 years living, studying, and working in Maharashtra, India, Curley returned to the USA to continue his studies and earned his Civil Engineering degree at Clarkson University, Potsdam, New York. For almost a decade, he focused on creating a watershed-wide, integrated flood control program for the Corte Madera Creek Watershed (Ross Valley), unifying the participating towns in a basin-scale approach to flood management, conducted successful campaigns for voter support for a 20-year revenue stream of $2.1 million/year, and completed a two-year feasibility study for over 180 measures to provide protection from the 1%

annual probability (i.e., 100-year) of flood. The measures include large detention basins, creek improvements, and sediment management in the channel. He created an ongoing stakeholder participation process including towns, advocate groups, environmental organizations, regulatory agencies, local utility districts, and interested citizens. Curley works as an independent consultant offering public outreach and communication services to local government agencies and businesses in central New Jersey, eastern Pennsylvania, and California.

Georges Descombes ADR Architects, Geneva. Descombes studied architecture in Geneva, Zurich, and London (AAGS dipl). He has taught at the University of Geneva, School of Architecture, and at the Berlage Institute, and served as a visiting professor at Harvard's Graduate School of Design, the University of Virginia, School of Architecture, the University of California Berkeley, and the Rapperswil School of Landscape Architecture. He has lectured in Europe, the USA, Israel, China, and South America. Among his better-known projects are the Park in Lancy, the Swiss Path, the Bijlmer memorial in Amsterdam, and the Parc de la Cour du Maroc in Paris. He is working on the revitalization of the River Aire in Geneva, on Lyon Confluence of the Soane and Rhone Rivers, and on the Ostende Green belt project in Belgium. The River Aire project has been awarded the Schulthess Swiss Gardens Prize 2012, the Audience Prize at the last Biennial of Landscape architecture in Barcelona, the Swiss Engineers Society Prize in 2017, and a Silver Medal at the International Prize for Sustainable Architecture in Ferrara (2017). Descombes was appointed a David Skinner lecturer at the Edinburgh School of Landscape architecture, a John R. Bracken Fellow in Landscape architecture at Pennsylvania State University, and a Regents Lecturer at the University of California Berkeley. In 2014 he received the Cultural Prize of the City of Geneva and in 2016 the Cultural Prize of the Leenaards Foundation in Lausanne.

Bill De Groot Urban Drainage and Flood Control District (retired). De Groot was the manager of the Floodplain Management Program for the Denver Urban Drainage and Flood Control District from February 1974 until his retirement. The program has been active in assisting local governments with floodplain regulations and the National Flood Insurance Program, delineation of 100-year floodplains ahead of development pressures, review of development proposals for conformance to district master plans and criteria, review of construction drawings for eligibility for district maintenance assistance, public information efforts, and flood detection and warning plans. De Groot has BS and MS degrees in Civil Engineering from South Dakota School of Mines and Technology. He is a registered professional engineer in Colorado. He was the director for the National Association of Flood & Stormwater Management Agencies (NAFSMA) for several years and is a member of the American Society of Civil Engineers and Association of State Floodplain Managers. He was a founding member of the Colorado Association of Stormwater and Floodplain Managers and its first secretary/treasurer. De Groot served as a technical advisor to the Federal Emergency Management Agency's Technical Mapping Advisory Council from 1997 to 2000.

Ralph Johnson Alameda County Flood Control and Water Conservation District (retired). Johnson graduated from UC Berkeley with a BS in Civil Engineering in 1970 and went to work for the Alameda County Flood Control District. He is a professional civil engineer and retired from the Flood Control District in 2000 as the deputy director of Public Works. While working for the Flood Control District, he worked on a large variety of projects, ranging from ground-water basin studies, land development, flood channel design, and sediment management. From 2003 through 2012, he represented the Flood Control District on the South Bay Salt Pond Restoration's Project Management Team and serves as a Board Member for the Castro Valley Sanitary District, the east Bay Dischargers Authority, the Alameda County local Agency Formation Commission, and the Alameda County Library Foundation.

Pilar Lopez-Llompart University of California Berkeley. Lopez-Llompart is a civil engineer specialized in hydrology and environmental engineering. She conducted her university studies at the Technical University of Catalonia (UPC-Barcelona Tech), Spain, and developed her master's thesis research at the Technical University of Denmark (DTU) (supported by an Erasmus grant) on the modeling of variable density flow in karstic coastal aquifers. She has worked on diverse projects, such as water supply infrastructure, urban drainage networks, flood risk assessment, and river restoration. As a visiting researcher at the University of California Berkeley (Center for Environmental Design Research), she conducted research on encroachments in two floodways of the Mississippi River.

David Mallory Denver Urban Drainage and Flood Control District (retired). Mallory served as a senior project engineer and manager of the Floodplain Management Program at the Urban Drainage and Flood Control District (UDFCD) in Denver, Colorado. Mallory is a graduate of Colorado State University and has nearly 40 years of experience in the Denver metro area, 20 years in the private sector, and 20 years with the UDFCD. He was first registered as a professional engineer in the state of Colorado in 1981. Mallory supervised all aspects of the program, including the review of development proposals adjacent to streams, floodplain mapping, and the Cooperating Technical Partnership (CTP) initiative with the Federal Emergency Management Agency (FEMA) including the letters of map change (LOMCs) delegation. He was selected to serve a two-year term on the Technical Mapping Advisory Council, which made a suite of far-reaching recommendations to FEMA to improve floodplain mapping, risk communication, and flood insurance delivery. Mallory is active in the Association of State Floodplain Managers (ASFPM), Colorado Association of Stormwater and Floodplain Managers (CASFM, past Chair), and the Natural Hazard Mitigation Association (NHMA). Mallory continues to contribute to floodplain management issues such as floodplain mapping, as well as emerging topics related to floodplain preservation and protection of the natural and beneficial stream and river functions.

Len Materman San Francisquito Creek Joint Powers Authority. Materman is the executive director of the San Francisquito Creek Joint Powers Authority (SFCJPA), a regional government agency named for the physical feature that divides and unites cities and counties in Silicon Valley. Since Materman joined the agency, its project funding has grown from zero to over $83 million. After decades of discussion among jurisdictions, the SFCJPA is now constructing a project to protect an underserved community beset by creek flooding and rising sea levels. During Materman's tenure, the agency has been highlighted as a model of regional collaboration in the *Wall Street Journal*, *Los Angeles Times*, and major Bay Area media. Previously, Materman served as the government affairs director at UC Berkeley, where he received two Chancellor's Distinguished Service Awards; as an adviser to the director of FEMA in Washington, DC, during the Clinton administration; as an executive producer of a film broadcast nationally on PBS; as an adviser to two interdisciplinary centers at Stanford University; and as a consultant to nonprofits, foundations, and the US State Department. Materman received degrees in political science and biological sciences from the University of California, Davis.

Rod Mayer HDR, Inc. Mayer is a senior technical adviser with HDR in Folsom, California. Prior to his retirement from the California Department of Water Resources (DWR) in 2013, he was DWR's FloodSAFE Executive. He has 40 years of experience in water resources planning, design, construction, operations, and maintenance. For the past 25 years he has worked on California's flood management programs, with an emphasis on Central Valley flood issues and projects. He is also a member of the National Committee on Levee Safety. He holds a master's degree in civil engineering and is a licensed civil engineer and geotechnical engineer in California.

Dale Morris Embassy of Netherlands, Washington. Morris is a senior economist at the Royal Netherlands Embassy in Washington, DC, where he directs the Dutch government's Water Management efforts in Louisiana, Florida, California, and Virginia. Morris focuses on broad "sustainability" topics: flood protection, flood risk mitigation, coastal and floodplain restoration, ecosystem resiliency, and urban water management and adaptation. He is a co-director of Dutch Dialogues, and a member of advisory boards and steering committees in New Orleans and Washington, DC. Morris previously worked in the US Congress, served in the US Air Force, and is a graduate of the University of Pittsburgh and the University of Virginia.

Pedro Pinto Instituto Superior Técnico, University of Lisbon. Pinto holds a PhD in Landscape Architecture and Environmental Planning, from the University of California, Berkeley. He received a degree in Territorial Engineering (equivalent to a M.Eng) and an MSc in Land-Use Planning, both from Instituto Superior Técnico (Lisbon, Portugal), where he has worked (2003–2009 and again since 2015) as a researcher. He has been involved in projects including school facility planning, the study of Portuguese cities with rivers, and municipal land-use planning at different scales. His master's thesis, "The Portuguese Fluvial City", included extensive spatial analysis of all 75 Portuguese cities with rivers. For his PhD dissertation, "Metropolitan Estuaries and Sea Level Rise", he simulated the impacts of sea level rise inundation of estuarine lowlands around the San Francisco Bay, California, USA, and the Tagus Estuary, Lisbon, Portugal; reconstructed the environmental histories of both estuaries resorting to ArcGIS; and analyzed environmental governance structures and readiness to face sea level rise.

Melisa Samet National Wildlife Federation. Samet is the senior water resources counsel for the National Wildlife Federation, where she directs and advises water resources campaigns focused on floodplain protection and improving the projects and programs of the Army Corps of Engineers. She has been instrumental in protecting hundreds of thousands of acres of wetlands and other sensitive habitats and in obtaining key water resources planning reforms. She has testified before Congress and the National Academies, authored numerous articles and reports on water resources policy, was appointed to a Clean Water Act Federal Advisory Committee, directed water resources and ocean protection campaigns for American Rivers and Earthjustice, and worked as a litigation attorney in private practice. She was awarded a 2009 National Wetlands Award from the Environmental Law Institute. Samet received her JD from the New York University School of Law and her BS in Wildlife Biology from the University of Vermont.

Graça Saraiva University of Lisbon (retired). Graça is a landscape architect and planner, and the former chair of the Department of Urbanism at the Faculty of Architecture (FA) of the University of Lisbon (UL), where she taught courses in landscape architecture, planning, and design until her retirement. She is trained as a landscape architect and agronomical engineer, with a master's degree in Urban and Regional Planning, and a PhD in Landscape Architecture, at the Institute of Agronomy, UL. Earlier a researcher at Centre of Urban and Regional System of the Instituto Superior Tecnico of UL, Graça researches at the Research Centre of Architecture, Urbanism and Design, of FA/UL. She has contributed to numerous European and national research projects on landscape planning, river restoration, flood management, natural resources planning, and landscape perception and sustainability. Concurrent with her academic duties, she also served, from 2005 to 2009, in the Portuguese Ministry of Environment as an adviser to the minister.

Laurent Schmitt University of Strasbourg. Schmitt is Professor of Physical Geography in the laboratory Image City Environment, Faculty of Geography and Planning, University of Strasbourg. A fluvial geomorphologist specialized in sustainable environmental river management and restoration, his research, developed in the frame of interdisciplinary projects, focuses on linking quantitative fluvial geomorphology to hydro-ecology. He also studies links between fluvial long-term trajectories and contemporary river dynamics, including restoration. He has conducted research on the Rhine, the Yzeron (periurban sub-catchment of the Rhône), the Nile, and several Italian rivers and deltas. He is the coordinator of three large monitoring programs concerning restoration actions in the Rhine Basin. He is the co-director of the Laboratory UMR 7362 CNRS and a member of the Research Committee of the University of Strasbourg, and was earlier a member of the National Comity of the CNRS.

Charles E. Shadie US Army Corps of Engineers, Mississippi Valley Division (retired). Schmitt was Chief of the Watershed Division for the Mississippi Valley Division, US Army Corps of Engineers, and the Mississippi River Commission from 2010 to 2016. He was responsible for overall supervision; policy formulation; and federal, state, and local entity coordination for regional operation for water management and hydrology, as well as hydraulic and coastal engineering within the Mississippi Valley Division (MVD). With a BS in civil engineering from West Virginia Institute of Technology, MS in civil engineering from Purdue, and MS in water resources planning and management from the University of Florida, he is a registered professional engineer in Louisiana and Mississippi. He started with the Chicago District in 1981 as a hydraulic engineer, working on urban flood control for the Chicago area, then in 1982 began his 14 years of service at New Orleans District, working on flood risk reduction, navigation, and environmental restoration, as project engineer and supervisory hydraulic engineer. At MVD, he has provided technical expertise to the Mississippi River & Tributaries Project, Louisiana Coastal Area study, and Louisiana Hurricane Recovery Projects.

Todd Strole Consultant. Strole was the associate director for Floodplain Management within the Nature Conservancy's (TNC) Mississippi River Basin Program, seeking innovative ways to restore floodplain function with both public and private funding and market-based approaches. He also coordinated TNC's Risk Reduction and Resilience Priority, promoting use of natural infrastructure within national policy and local planning. Based in St. Louis, he works on floodplain issues throughout the country, with TNC and other entities. Strole has extensive experience with the Corps of Engineers, including project planning under several Corp authorities. He completed a three-year work detail from 2009 to 2011 with the Corps of Engineers under an Inter-governmental Personnel Act agreement, where he worked on floodplain restoration issues in the Upper Mississippi River, coordinating activities among three Corps Districts and including participation from stakeholder groups and academia. Strole is an advisory board member for the National Great River Research and Education Center in Alton, IL, and past chair of the Middle Mississippi River Partnership. Strole has BS degrees in environmental biology and botany from Eastern Illinois University and an MS degree in biological sciences from Illinois State University.

Timothy Washburn Sacramento Area Flood Control Agency. Washburn served as the chief counsel for the Sacramento Area Flood Control Agency (SAFCA) from 1990 to 2009 when he was appointed the agency's director of planning. Washburn has represented SAFCA in all aspects of project planning, finance, legislative advocacy, land acquisition, and environmental compliance. Before joining SAFCA, he served for two years as a Deputy City Attorney for the City of Sacramento, following two years in private practice with the firm of Weintraub, Genshlea, Hardy, Erich & Brown in Sacramento. He graduated as a history major from the University of California in 1983 and received his law degree from the University of California, Davis, in 1986.

David L. Wegner Jacobs Engineering. Until 2015, Wegner was a senior Democratic staff for the Transportation and Infrastructure Committee, US House of Representatives, with an oversight of the Army Corps of Engineers, water programs at the US Environmental Protection Agency (EPA), Tennessee Valley Authority, portions of the NRCS watershed programs, and tribal water issues. He earned a bachelor's of science from the University of Minnesota, and a master's from Colorado State University in river engineering. He worked for the US Interior Department for more than 20 years, 14 of those coordinating the science program in the Grand Canyon. Prior to his appointment to committee staff in 2009, he ran a consultancy focused on ecological issues related to dams globally. Wegner works as a part-time senior scientist with Jacobs Engineering, continuing to provide input and strategic counsel to Members of Congress, the US Bureau of Reclamation, Army Corps of Engineers, and international organizations. He serves on the National Academy of Sciences Water Science Technology Board and the National Reservoir Sedimentation Team.

Raymond Wong GHD. Wong is a water resource engineer with GHD, specializing in urban watershed planning, flood management, and restoration. In his professional practice, Wong oversees studies and design projects throughout California, with technical emphasis on hydraulic, hydrology, geomorphology, modeling, and urban streams and infrastructure design including LID. Wong is also a consulting project manager at the City of Mountain View, advising the city on watershed flood management, sea level rise adaptation, and tidal marsh restoration in local and regional scales. Wong has a PhD in environmental planning from the University of California Berkeley, with a research focus on management issues with structural flood protection in urban streams, particularly on aging federal flood control projects. Wong also has an MS in environmental engineering from Stanford University and a B.A.Sc. Honors in civil engineering from the University of Toronto.

Liang Xu Santa Clara Valley Water District. Xu has a PhD from Arizona State University and has over 20 years of experience in hydraulics, sediment transport, and planning and design of flood protection projects in the USA and China. He is an engineering unit manager at the Santa Clara Valley Water District. He is responsible for the development of computer programs for hydraulic analysis and geomorphology studies. He also supervises the hydraulic engineering analyses for flood protection projects and works on designs for geomorphically stable channel design techniques for flood protection projects.

LIST OF FIGURES

CHAPTER 1

Introduction

Anna Serra-Llobet, G. Mathias Kondolf,
Kathleen Schaefer, and Scott Nicholson

Abstract Managing flood risk instead of 'controlling floods' is a key change in approach for managing floods and floodplains. In the context of floods, hazard refers to the magnitude or height of a given flood and its probability of occurring. Vulnerability refers to the social assets exposed to damage from flooding. Risk combines both hazard and vulnerability. Thus managing flood risk implies making interventions at all points of the flood risk

A. Serra-Llobet (✉)
University of California Berkeley, Berkeley, CA, USA

Aix-Marseille University, Marseille, France

G. M. Kondolf
University of California Berkeley, Berkeley, CA, USA

University of Lyon, Lyon, France

K. Schaefer
University of California Berkeley, Berkeley, CA, USA

Federal Emergency Management Agency, Oakland, CA, USA

S. Nicholson
University of California Berkeley, Berkeley, CA, USA

US Army Corps of Engineers, Washington, DC, USA

© The Author(s) 2018
A. Serra-Llobet et al. (eds.), *Managing Flood Risk*,
https://doi.org/10.1007/978-3-319-71673-2_1

1

cycle, including not only structural measures to reduce flood magnitude or frequency (reducing hazard) but also land-use planning (to reduce assets exposed), early warning systems, insurance, and acting within the context of multiple objectives. Recent experiences in implementing flood risk management along large floodplain rivers and smaller urban streams in America and Europe manifest a wide range of environmental and institutional settings, and thus opportunities and constraints unique to each setting.

Keywords Flood risk management • Flood control • Large floodplain rivers • Urban rivers

Human societies have long settled along rivers and for just as long have been dealing with flooding. Today flooding remains by far the biggest and costliest natural hazard globally. Early societies accommodated annual and less frequent cycles of flooding through locating settlements and infrastructure on higher ground where possible, raising structures, and seasonal occupation of lowlands. Diverting floodwaters away from cities through bypass routes or out-of-basin diversion was practiced by the ancient Nabateans (Mays 2010) and by the Romans, who debated the relative impacts of diverting flood flows of the Tiber into neighboring basins (Keenan-Jones 2013). As technologies evolved, larger-scale structural methods of control were employed, leading to the massive dams built from the mid-twentieth century onward, some of which have included flood control among their objectives and extensive systems of dikes to limit flooding.

The past half century has seen an evolution in thinking from *flood control* to *flood risk management*, reflecting increased understanding that building structures to control floods is only one of many possible approaches available to societies (Sayers et al. 2013). Understanding floods as a process, and not as isolated events in time and space, we can distinguish four phases for flood risk management in the framework of the flood risk management cycle: (1) the characterization of hazard and risk (assessment and mapping); (2) mitigation strategies, which include prevention measures (e.g., land-use management) and protection measures (e.g., levees and dams); (3) emergency management, meaning preparation and response; and (4) recovery at short and long terms (Serra-Llobet et al. 2016) (Fig. 1.1).

Before going much further, it may be worthwhile to define 'risk' and distinguish the terms from related terms such as hazard, exposure, and vulnerability. As noted by prior authors (e.g., Merz et al. 2010; Gallopin 2006), these terms have been variously used and defined. Here we use

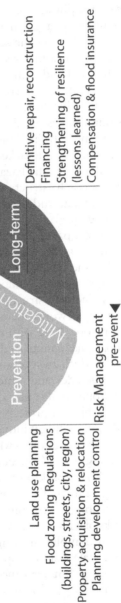

Fig. 1.1 The flood risk management cycle (Source: Image adapted from Serra-Llobet et al. 2016)

hazard for the physical process and its likelihood of occurring, for example, a flood reaching a given elevation with a given probability each year. *Vulnerability* denotes the societal assets exposed to flooding and accounts for the socio-economic system to recover from a flood. *Risk* results from both the hazard and vulnerability. If there are no people, infrastructure or other assets that we value on the floodplain to be damaged, there is no risk, no matter how big the floods are (i.e., no matter how high the hazard).

By siting villages on high ground above the floodplain, as was the case along the Upper Rhine Valley (Plate 2002), traditional societies reduced their vulnerability but lacked any means to reduce the frequency or magnitude of flooding (the hazard). However, in the twentieth-century 'flood control' approach, structural measures were implemented to reduce the hazard (i.e., magnitude and frequency of flooding). These structural measures (e.g., dams to reduce the flood flow in the river, levees to keep flood-waters out of a part of the natural floodplain) commonly induced further settlement in floodplains that were still exposed to flooding (though less frequently), thereby increasing vulnerability (Tobin 1995).

Moreover, the conventional structural measures were typically single-purpose engineering structures, with negative environmental consequences. In the developed world, increased requirements for environmental protection and restoration, along with recognition of global change and its effects on water supply, have motivated adoption of *integrated flood risk management* (IFRM), in which measures to manage floods must be developed and analyzed in a broader and "integrated" context, and multi-objective projects (with environmental, water quality, and recreational benefits) are preferred over single-objective projects. While there is increasing agreement that IFRM is a good idea, putting it into practice remains a challenge. Diverse views have been advanced regarding which governance systems are better suited to implementing IFRM (e.g., bottom-up approaches as in the USA or top-down as in the EU) (Serra-Llobet et al. 2016). Floods have a geographical extent, but managing them is done in the context of administrative, social, and cultural boundaries, which commonly do not align. The regulatory and political frameworks of a region determine to some extent what options are feasible.

IFRM also implies moving from a site- or project-based approach to a catchment-scale systems approach, which means not only relying on structural measures but also incorporating non-structural measures such as land-use planning, flood mapping, early warning, and controlled flooding for ecological benefits.

The options available to manage flood risk can be seen in a framework of temporal and spatial scales, as illustrated in Fig. 1.2. For example, actions

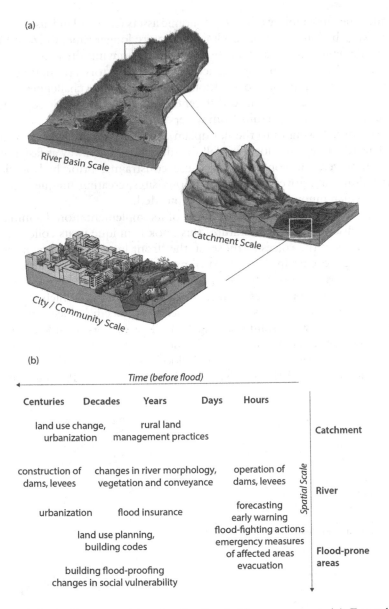

Fig. 1.2 Different spatial scales for flood risk management (**a**) Examples of human effects on flood risk and flood risk management measures at different temporal and spatial scales (**b**) (Source: (**b**) table adapted from Merz et al. 2010)

to reduce the vulnerability of socio-economic assets (such as land-use planning to keep buildings out of floodplains) require longer time horizons to affect settlement patterns and decisions on siting large infrastructure.

Increasingly, we see public agencies and other actors attempting to implement IFRM in the face of multiple challenges. These challenges vary with context and can be quite different along major river floodplains (where the flood hazard results from waters conveyed from upper parts of the basin, overflowing onto the floodplain) and urban areas (where most flood hazards result from local rainfall that does not drain away as fast as it falls). Moreover, the context of land-use constraints, prior hydrologic modifications, and governance varies among sites, creating unique situations for which innovative approaches are needed.

While the concept of IFRM and some of its implementation dilemmas have been discussed in the literature, this book is unique in its collection of experiences in the voices of those on the 'front lines' of implementing integrated approaches in the context of multiple constraints, the greatest of which are usually institutional. The contributions in this book come from a wide range of river basins in America and Europe, in which we can see how the specific settings both influenced the specific objectives and constrained the possible solutions (Fig. 1.3). The authors include a wide range of scholars and practitioners, with a dominance of the latter: reporting their direct experiences dealing with flood risk, problems that remain (such as limitations of the US National Flood Insurance Program and

Fig. 1.3   Geographical location of the chapters

ways to work around these), and some of the innovative approaches that have resulted (such as the plan adopted by Contra Costa County, California, to convert deteriorating concrete channels into natural channels over the next 50 years).

The book is divided into two parts. The first focuses on big river basins, the second on urban streams. As we see from the specific case studies, different spatial scales require different strategies and measures to manage floods, and of course, context is everything. Of North American rivers, the Mississippi and Sacramento have arguably been the most influential in creating and modifying flood management policy because of the early settlement pressure in these areas and the economic importance of activities directly affecting the rivers. It is on the Mississippi and Sacramento where the US Army Corps of Engineers got its first significant experience in managing floods and where some of the limitations of flood risk management in the US style, through the National Flood Insurance Program, became apparent. While the Mississippi is much larger, the Sacramento is still a major river system characterized by broad floodplains and altered hydrology on a catchment scale. Both river basins are now highly engineered, with flood control systems that depend on a combination of reservoirs, levees, and flood bypasses (Chaps. 2 and 3). The Rhine has had outsized importance in European history and culture. Also highly engineered, the Rhine is being transformed by multiple floodplain reconnection projects that aim to reduce flood risk and restore ecosystem functions, both in the upper Rhine and the delta (Chap. 4).

At the urban scale, past efforts to 'control' flood through structural measures have left a legacy of highly engineered channels, which are increasingly recognized as unsustainable by virtue of having been built to undersized standards, whose performance has been less even than designed, and which have had unacceptable environmental and social consequences. These problems are well illustrated around San Francisco Bay, where many cities are trying to complement hard hydraulic infrastructures with non-structural measures, and increasingly to view options in a larger catchment context. The most difficult problem is that of already-urbanized lands in floodplains and along the shore of the estuary, a problem much less severe in the estuary of the Tagus River, Lisbon, where the central government historically prevented urbanization of flood-prone lands. Thus, very different experiences are reported by flood control districts in the San Francisco region and California and those for managing flood risk in urban streams of the Lisbon area, despite their comparable spatial scales, orographic features, and Mediterranean climates. The differences are due both to historical governance systems and to recent EU-wide policy that requires systematic assessment of flood risk (i.e., not only hazard but also vulnerability) and development of measures to reduce risk. Chap. 5 presents case studies of how various agencies in these two Mediterranean-climate regions have tried to diversify their flood risk management strategies by incorporating non-structural measures and also to integrate ecological values. In the mid-continental, urban settings of Denver (Colorado) and Geneva, Chap. 6 explores the evolution of flood control strategies in the greater Denver area (an unusually effective catchment-scale, integrated flood management program) and a landmark project on the Aire River of Geneva that provides both flood risk management and ecological benefits through an *espace de liberté* approach.

The book concludes with some reflections on the diverse experiences reported by the contributors and the lessons that can be drawn from them (Chap. 7).

REFERENCES

Gallopin, G.C. 2006. Linkages Between Vulnerability, Resilience, and Adaptive Capacity. *Global Environmental Change* 16: 293–303.

Keenan-Jones, D. 2013. Large-Scale Water Management Projects in Central-Southern Italy. In *The Ancient Mediterranean Environment Between Science and History*, ed. W.V. Harris, 233–258. Leiden: Brill.

Mays, L.W. 2010. A Brief History of Water Technology During Antiquity: Before the Romans. In *Ancient Water Technologies*, ed. L.W. May, 1–28. Dordrecht/London: Springer Science and Business Media.

Merz, B., J. Hall, M. Disse, and A. Schumann. 2010. Fluvial Flood Risk Management in a Changing World. *Natural Hazards and Earth Systems Sciences* 10: 509–527.

Plate, E.J. 2002. Flood Risk and Flood Management. *Journal of Hydrology* 267: 2–11.

Sayers, P., Y. Li, G. Galloway, E. Penning-Rowsell, F. Shen, K. Wen, Y. Chen, and T. Le Quesne. 2013. *Flood Risk Management: A Strategic Approach*. Paris: UNESCO.

Serra-Llobet, A., E. Conrad, and K. Schaefer. 2016. Governing for Integrated Water and Flood Risk Management: Comparing Top-Down and Bottom-Up Approaches in Spain and California. *Water* 8: 445. https://doi.org/10.3390/w8100445.

Tobin, G.A. 1995. The Levee Love Affair: A Stormy Relationship. *Water Resources Bulletin* 31: 359–367.

PART I

Big River Basins

CHAPTER 2

Managing Floods in Large River Basins in the USA: The Mississippi River

Charles E. Shadie, Pilar Lopez-Llompart, Melissa Samet,
Todd Strole, and G. Mathias Kondolf

Abstract The Mississippi River was the first theater in which the federal government sought to control floods and improve navigation through the efforts of the US Army Corps of Engineers, initially under a "levees only" philosophy, later revised (after the disastrous 1927 flood) to include multiple approaches, such as backwater areas and flood bypasses. The Mississippi River and Tributaries Project successfully conveyed the 2011 flood (with more rainfall than fell in 1927), but operation of critical

C. E. Shadie
Mississippi Valley Division, US Army Corps of Engineers, Vicksburg, MS, USA

P. Lopez-Llompart
University of California Berkeley, Berkeley, CA, USA

M. Samet
National Wildlife Federation, Washington, DC, USA

T. Strole
Consultant, St Louis, MS, USA

G. M. Kondolf (✉)
University of California Berkeley, Berkeley, CA, USA

University of Lyon, Lyon, France

A. Serra-Llobet et al. (eds.), *Managing Flood Risk*,
https://doi.org/10.1007/978-3-319-71673-2_2

11

bypasses was threatened by encroachment of buildings within the bypasses, permitted by local governments. Structures designed to concentrate flow for the benefit of navigation can result in higher flood stages and thus can undermine flood control efforts. Allowing floodplains to flood naturally, as much as possible, can have benefits not only for the ecosystem but also for managing floods to minimize inundation of cities.

Keywords Mississippi River • Mississippi River 2011 flood • New Madrid Floodway • West Atchafalaya Floodway • Effects of navigation structures • Floodplain benefits

2.1   INTRODUCTION

Charles E. Shadie and G. Mathias Kondolf

2.1.1   The Mississippi River Basin

The Mississippi River Basin drains 41% of the 48 contiguous states of the USA. Its 3.2-million km² basin (1.25 million mi²) extends from the Rocky Mountains to the Appalachian Mountains, the largest river system in North America and the third-largest river basin in the world (Fig. 2.1). Its basin roughly resembles a funnel with its spout at the Gulf of Mexico. The lower alluvial valley of the Mississippi River is a relatively flat plain of about 90,600 km² (35,000 mi²), which historically flooded during times of high water prior to the construction of flood protective works, which were begun in the late 1700s. The Mississippi River system has an average flow into the Gulf of Mexico (via the Mississippi and Atchafalaya Rivers) of about 18,100 m³/s (640,000 ft³/s), much greater during floods. In one of the largest floods recorded on the Mississippi, in May–June 2011, the peak flow into the Gulf of Mexico was over 68,000 m³/s (2.4 million ft³/s). The Mississippi River has been extensively altered for navigation and flood control (Alexander et al. 2012).

The 2011 flood brought into focus many issues in the Mississippi River Basin, such as the role of the federal flood control project in this inter-state basin, the resistance of local interests to honoring flowage easements on their properties, and the conflict between structural approaches to flood

Fig. 2.1 The Mississippi River Basin

control and ecosystem services provided by naturally functioning flood-plains. In this chapter, Charles Shadie of the US Army Corps of Engineers (Corps) points out that during the 2011 flood, the mid-twentieth-century flood control project worked largely as planned, preventing an estimated over $110 billion in flood damages (Sect. 2.2). In this section, he also notes that some navigation features of the Mississippi River and Tributaries (MR&T) Project (channel cutoffs, channel dredging, etc.) reduced the severity of flooding. The role of levees and navigation structures on flood heights has been contested for half a century, including Belt's (1975) conclusion that the record flood stages in 1973 were "manmade" due to "the combination of navigation works and levees." Melissa Samet considers the conflicts inherent between the Corps' objectives of navigation and flood risk management on the Mississippi, and argues that navigation works have increased flood risk (Sect. 2.4). Pilar Lopez-Llompart and Matt Kondolf consider land-use conflicts arising in the federally designated floodways (flood bypasses), where local jurisdictions have given building permits for structures within the floodways themselves, creating inevitable conflicts with the designated uses of the lands within the bypasses, not only exposing houses to flooding, but also compromising operation of the floodways (Sect. 2.3). Notably, in 2011 the state of Missouri sued to prevent a floodway from being activated, but was turned down by the Supreme Court just in time for the Corps to activate the floodway and avoid flood damages to settlements elsewhere. Todd Strole summarizes the benefits of connected floodplains and some successful efforts to restore these ecosystem functions along the highly altered Mississippi River system (Sect. 2.5).

2.1.2   Flood Risk Management in the Mississippi: History and Governance

With its major tributaries the Missouri, Ohio, Arkansas-White, and the Red Rivers, the Mississippi River drains all or part of 31 states and two Canadian provinces (Fig. 2.1). Given the multiple states in the basin, the potential for conflicting priorities among them, and the value of the assets exposed to flooding, the need for a federal role in controlling floods was obvious.

The 1927 Mississippi River flood was the greatest natural disaster up to that point in the US history, as many levees overtopped and breached, and between 120 and 225 crevasses developing, 17 of those being major crevasses on federal levees. The remainder of the breaks—ranging in size from half a mile wide to a mere trickle—occurred in state or local levees. By the time the flood finally subsided in August 1927, over 67,000 km² (26,000 mi²) or 72% of the Lower Mississippi Valley had been inundated to depths up to 9.1 m (30 feet), levees were crevassed, and cities, towns and farms lay in waste. Where it reclaimed the floodplains the river was now in some places up to 160 km (100 miles) across. Crops were destroyed and industries and transportation paralyzed. The human loss was staggering as well, with up to 250 people killed directly, and deaths due to disease and exposure after the flood likely exceeding 1000. In addition, about 162,000 homes were unlivable, and 41,000 buildings were destroyed resulting in over 600,000 people being left homeless, with many having to live in tents for months following the flood. At a time when the federal budget barely exceeded $3 billion dollars, the flood, directly and indirectly, caused an estimated $1 billion dollars in property damage.

In response to the 1927 flood, the US Congress passed the 1928 Flood Control Act, authorizing the MR&T project, which represented one of the first comprehensive public works projects within the Lower Mississippi Valley that would provide enhanced protection from floods while maintaining a mutually compatible and efficient Mississippi River channel for navigation. The project also represented a major departure from relying solely on levees for flood protection. Prior to 1927, local, state, and eventually federal agencies provided flood protection via a "levees only" approach, building levees higher and higher after larger floods overtopped the existing lines of protection. However, the 1927 flood demonstrated that "levees only" would not be adequate to provide the level of flood protection needed along the Lower Mississippi River Valley. The plan developed by the USACE Chief Engineer General Edgar Jadwin would provide flood protection for floods larger than the 1927 flood by acknowledging that, for some floods, measures in addition to levees would be required (Jadwin 1928; Barry 1997).

The MR&T Project consists of four primary components: (1) an extensive levee system to prevent overflows on developed alluvial lands, with a total of 6000 km (3727 miles) of mainstem and tributary levees and floodwalls that were authorized for the MR&T system, of which about 5600 km (3486 miles) have been built; (2) floodways and backwater areas to safely divert or store excess flows at critical reaches so that the levee system will not be unduly stressed. There are four floodways: one in Missouri (Birds Point—New Madrid Floodway) and three in Louisiana (Morganza Floodway, Bonnet Carré Spillway, and the West Atchafalaya Floodway) (see Sect. 2.3), and four backwater areas: St. Francis Backwater Area in Missouri and Arkansas, White River Backwater Area in Arkansas, Yazoo Backwater Area in Mississippi, and Red River Backwater Area in Louisiana (Fig. 2.2); (3) channel improvements and stabilization features (such as meander cutoffs, bank and channel revetments, channel bendway weirs and stone dikes, and dredging) to protect the integrity of flood control features and ensure proper alignment and depth of the navigation channel; and (4) tributary basin improvements, including levees, headwater reservoirs, and pumping stations, designed to expand flood protection and improve drainage into adjacent areas within the alluvial valley (Davis et al. 2017).

2.2   THE 2011 MISSISSIPPI RIVER FLOOD: WHAT WORKED

Charles E. Shadie

The 2011 flood in the Lower Mississippi River produced record flows throughout the 90,600 km² (35,000 mi²) river basin and resulted in record or near-record stages throughout the lower valley (Camillo 2012). The 2011 flood provides an opportunity to assess the effectiveness of the system of flood control structures in place in the region (Shadie and Kleiss 2012; Davis et al. 2017).

The 2011 flood set records for flow and stage over much of the Lower Mississippi River Basin, testing the MR&T Project as never before. Levees

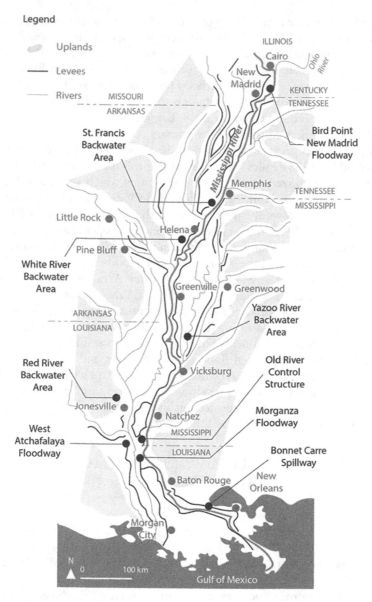

Fig. 2.2 Location of features of the Mississippi River and Tributaries (MR&T) Project. In addition to floodways, backwater areas and principal levees are shown (Source: Redrawn from US Army Corps of Engineers, Davis et al. 2017)

and floodwalls throughout the project experienced higher stages and pressures than from previous floods. In some areas, stages threatened to overtop the levees and floodwalls, requiring the USACE and local emergency crews to flood fight those areas with earthen berms, sandbags, and/or HESCO bastions (rectangular wire mesh containers) to prevent overtopping. While the levees and floodwalls held, hundreds of sand boils occurred throughout the system requiring emergency measures to stabilize the boils and prevent undermining and failure of the levee system. Sand boils had occurred before, throughout the MR&T system, in previous floods, but the 2011 flood placed higher pressures on the system (Shadie and Kleiss 2012).

Floodways played a larger role during the 2011 flood than ever before. For the first time since the project's inception in 1928, a total of three of the four MR&T floodways were operated during a flood. The first floodway operated was the Birds Point—New Madrid Floodway (Missouri). On May 2, 2011, the US Army Corps of Engineers detonated explosives placed in the Birds Point—New Madrid Floodway fuseplug levee to open the floodway and reduce stages and pressures along the levee system. With a record flow of at least 59,500 m³/s (2.1 million ft³/s) in the river, a peak of about 11,300 m³/s (400,000 ft³/s) was diverted away from the river and down the floodway, providing floodplain storage of 525 km² (130,000 acres) with depths up to 6 m (20 feet) (Davis et al. 2017).

As the flood continued to flow down the Lower Mississippi Valley, new record flows and stages were set threatening the levee system in many areas. At Vicksburg, Mississippi, the new record stage of 31.5 m NGVD29 (103.3 feet NGVD29) came within 7–10 cm (3–4 inches) of overtopping the Yazoo Backwater levee, almost placing that backwater area into operation. Further south, the Bonnet Carré Spillway, about 48 km (30 miles) upstream of New Orleans, was opened on May 9, 2011, diverting flows away from the river into Lake Pontchartrain eventually reaching a peak diversion rate of about 8920 m³/s (316,000 ft³/s).

Finally, on May 14, the Morganza Floodway structure, about 64 km (45 miles) upstream of Baton Rouge, Louisiana, was opened (Fig. 2.3). At its peak, the structure diverted 5150 m³/s (182,000 ft³/s) from the river into the Atchafalaya Basin. This was only the second time the Morganza Floodway had ever been operated. The West Atchafalaya Floodway, with a design flow of 7080 m³/s (250,000 ft³/s), was not operated, as its fuseplug (i.e., control weir) did not overtop, both because the Red/Ouachita rivers were not flooding and because the Atchafalaya River channel had downcut in the preceding decades, meaning that a greater flow is now

Fig. 2.3 Morganza Floodway in operation during the 2011 Mississippi flood (Source: US Army Corps of Engineers)

needed to overtop the fuseplug (see Sect. 2.3). Of a total of 148,000 ha (366,000 acres) in the four MR&T floodways, 85,800 ha (212,000 acres) were flooded during this event while the three floodways were operated.

None of the four MR&T backwater areas were operated during the 2011 flood although the Yazoo Backwater levee came close to overtopping. However, because of high Mississippi River stages, the drainage structures in those backwater areas had to be closed. As a result, some flooding in those backwater areas occurred from internal runoff. Of a total 669,000 ha (1,652,000 acres) in the backwater areas, 135,600 ha (335,000 acres) experienced some flooding. However, the backwater areas clearly had excess capacity for a flood of an even greater magnitude than the 2011 flood (Shadie and Kleiss 2012; Davis et al. 2017).

Navigation improvements undertaken as part of the MR&T Project reduced water levels during the 2011 flood event as well. From 1933 to 1942, a total of 15 meander bends were artificially cut off (and an additional natural cutoff occurred). These cutoffs along with dredging chute enlargements and other modifications shortened the river by over 270 km (170 miles) between Memphis, Tennessee, and Baton Rouge, Louisiana. While the 2011 flood set new stage records from Cairo, Illinois, to Caruthersville, Missouri, and from Vicksburg, Mississippi, to Red River Landing, Louisiana, by 30–60 cm (1–2 feet), the middle reach from Memphis, Tennessee, to Greenville, Mississippi, ranged from about 0.6 to

1.8 m (2–6 feet) below previous records from 1927 or 1937. The cutoffs completed from 1933 to 1942 are primarily responsible for this middle reach not setting new records in 2011. In addition, other channel control features (dikes, bendway weirs, revetments, etc.) contributed to stabilizing the channel and protecting the levees, thereby helping the project perform as intended.

Tributary basin improvements also provided flood risk reduction benefits during the 2011 flood. Features such as the five MR&T-authorized reservoirs and the St. Francis Basin Huxtable Pumping Station (capacity 340 m³/s (12,000 ft³/s)) stored or evacuated flood waters, reducing flooding of interior areas.

By the time the flood subsided in late June 2011, over 25,640 km² (9900 mi²) had been inundated. However, none of the project levees were breached or overtopped during the event (other than the Birds Point—New Madrid fuseplug levees which were detonated to activate the floodway). Even more important and remarkable, no deaths attributable to the flood occurred despite the fact that over 4 million people live and work within the Lower Mississippi Valley floodplains.

Since its initiation, the MR&T program has brought an unprecedented degree of flood protection to the project area within the Lower Mississippi Valley. The federal government contributed about $14.0 billion toward the planning, construction, operation, and maintenance of the project. The MR&T Project has provided a 44-to-1 return on that investment, including over $612 billion in flood damages prevented (including an estimated value of over $110 billion in 2011 alone), and waterborne commerce increases from 30 million tons in 1940 to nearly 500 million tons today. These figures place the MR&T Project among the most successful and cost-effective public works projects in the history of the USA.

2.3 Land-Use Conflicts in Floodways of the Mississippi River System

Pilar Lopez-Llompart and G. Mathias Kondolf

2.3.1 Introduction

Many national policies must be implemented by state and local governments, which have different motivations and constraints than the national government (May and Williams 1986). Local governments have primary responsibility for land-use planning, and many have permitted proliferation of development on flood-prone lands, in conflict with national policies, because they have "little fiscal stake...[and]... few incentives ...to be fully involved in floodplain management" (Galloway 1995:11).

The National Flood Insurance Program (NFIP), authorized by the US Congress in 1968, provided federally subsidized flood insurance for residents of floodplains, effectively a "...'carrot-and-stick' philosophy – making federal benefits contingent upon local zoning..." (Houck 1985:78). However, the divergent motivations of local governments can undermine effective implementation of the program, resulting in further encroachments of housing and infrastructure into flood-prone areas. This "implementation dilemma" (May and Williams 1986) is brought into sharp focus in the land-use history of nationally designated floodways along the Mississippi River (Kondolf and Lopez-Llompart 2018).

As described in Sect. 2.1, the MR&T Project included four designated flood bypasses (termed "floodways" in MR&T parlance), areas of floodplain designated to accommodate part of the river's flood flow, thereby reducing stage in the main river (Fig. 2.2). Since the initial planning of the MR&T, the Birds Point—New Madrid (New Madrid) and West Atchafalaya Floodways were opposed by residents who did not want their properties flooded to protect other lands along the valley (MRC 2007a). The US Army Corps of Engineers purchased flowage easements from the owners of all the affected private properties. However, the easements included no restrictions on the use or development of the land, and local jurisdictions have permitted many structures in these floodways.

2.3.2   *The Birds Point: New Madrid Floodway*

Completed in 1932, the New Madrid Floodway is designed to divert flood flows from the mainstem Mississippi River, thereby reducing the river stage and preventing overtopping of levees elsewhere. It is activated by blasting a breach in the levee at the upper end of the floodway when a peak stage of 18.3 m (60 feet) is forecast for Cairo, Illinois, to produce a decrease in stage on the Mississippi and Ohio rivers along the east bank opposite the floodway (MRC 2007b).

Despite the US government's flowage easements over all the lands within the floodways, during the record flood of 2011, activation of the New Madrid Floodway by the national government (US Army Corps of Engineers) was delayed by a lawsuit brought by the State of Missouri attempting to prevent inundation of lands within the floodway. Lower courts and finally the US Supreme Court rejected the suit and confirmed that the floodway should be operated as established in the MR&T Project (Camillo 2012). On May 2, 2011, the levee was detonated and water diverted through the floodway (Olson and Morton 2012a; Londoño and Hart 2013), lowering river stage at Cairo and elsewhere along the east bank of the river (Luke et al. 2015; Olson and Morton 2012b).

At the floodway's downstream end, a 460-m (1509 feet) gap in the levees allows floodwaters to return to the main Mississippi channel, and during smaller floods when the bypass is not activated, the gap allows backwater flooding from the Mississippi River to inundate the lowest one third of the floodway, providing shallowly flooded habitat of high value to fish and other wildlife (MRC 2007b). Agricultural interests have long called for this gap to be closed to prevent the backwater flooding and thereby permit farming in the floodway during high flows. However, the inundated floodplain habitat that exists now (and which would be lost if the gap were closed) is the kind of habitat now widely recognized as critically important for riverine ecosystems (Opperman et al. 2009; Dorothy and Nunnally 2015). The St. Johns-New Madrid Floodway project, originally authorized in 1954 to close the gap, was finally started in 2003 but was halted by a federal court ruling that the project had violated the Administrative Procedure, Clean Water, and National Environmental Policy Acts (Taylor 2007; Morton and Olson 2013; USACE 2015). Continued pressure for the project from local interests in Missouri (Dorothy and Nunnally 2015) met strong objections from the conservation community and from elected official representing residents along the river (on the opposite bank and upstream, in other states) whose risk of flooding would be increased (Barker 2014). In a multi-agency decision issued in January 2017, the US Army Corps agreed not to proceed with the project unless the project's impacts could be fully mitigated through advance restoration of a comparable area of frequently inundated floodplain, effectively meaning that to close off this connected floodplain area the Corps would have to open a comparable floodplain area to frequent flooding elsewhere (Wittenberg 2017).

2.3.3   The Atchafalaya Floodway System

The Atchafalaya River is the principal distributary channel of the Mississippi. From its bifurcation at the Old River Control Structure, the Atchafalaya flows westward, is joined by the Red River, turns southward, paralleled by and then receiving discharge from the West Atchafalaya Floodway as well as from the Morganza Spillway (from the mainstem Mississippi). Downstream, the combined floodway is termed the Atchafalaya Basin Floodway, ultimately discharging into the Gulf of Mexico (USACE 1938). The 69-km-long (42.9 miles) West Atchafalaya Floodway covers a surface of 610 km^2 (235.5 mi^2), mostly swampland, separated from the Atchafalaya River by a levee. The floodway was designed to be activated at 19,300 m^3/s (681,600 ft^3/s) by passive overtopping of the levee's northern end (MRC 2007a; FEMA 1980), to lower stages in the Atchafalaya and Red rivers, and in the Mississippi River itself. Under the MR&T plan, this floodway is the last component of the MR&T system to be activated (MRC 2007b), and in fact, it has never been used.

The Atchafalaya River channel has incised in recent decades, attributed to increased and sediment-starved flows (due to the water diverted into the Atchafalaya from the control structure having disproportionately lower sediment loads), and the effects of river engineering such as dredging, channel straightening, revetments, and wing dikes (Mossa 2016). Due to the increased capacity of the Atchafalaya River from channel incision, a much larger flow is probably needed now to passively overtop the fuseplug levee section and initiate flow through the West Atchafalaya Floodway.

Describing the situation in the early days of the NFIP, Houck (1985) documented extensive building within designated floodways of the Atchafalaya Floodway system. One community lay half within the Atchafalaya Basin Floodway, but local officials were "reluctant to limit growth in so large an area." In Point Coupee Parish, which includes the Atchafalaya River itself east of the West Atchafalaya Floodway, the Federal Emergency Management Agency found local official had allowed extensive development, evincing a "…'total lack of understanding' of the NFIP program, and gross neglect of FEMA's regulations" (Houck 1985:99).

To document recent trends in land use within the West Atchafalaya Floodway, Lopez-Llompart and Kondolf (2016) mapped the buildings and other structures (which are considered *encroachments* within the designated path of floodwaters) and found that the number had tripled (from 1439 to 4324) from 1968–1969 to 2008–2009, mostly after 1994 (Fig. 2.4 a–c). The highest density occurred around the town of Simmesport, which lies outside of the floodway within a ring of levees, but whose growth has "spilled over" into the floodway itself.

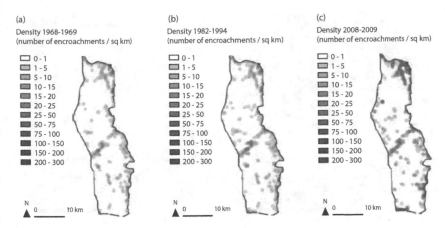

Fig. 2.4 Density of encroachments (number of encroachments/km²) in West Atchafalaya Floodway at the three studied time periods: (a) 1968–1969, (b) 1982–1994, and (c) 2008–2009 (Source: Lopez-Llompart and Kondolf 2016)

While the widespread construction within flood-prone areas is not unique, reflecting as it does a lack of enthusiasm by local governments to enforce land-use restrictions associated with the federal flood insurance program, the encroachments within the floodways have implications that go beyond inundation of the poorly sited structures themselves. Although the structures and their parcels occupy less than 2% of the total area of the West Atchafalaya Floodway, the encroachments concentrate along east-west trending roads traversing the floodway normal to the flow direction (Fig. 2.5). This linear pattern may have implications for hydraulic roughness during a flood, potentially decreasing the conveyance of the floodway during large floods.

2.3.4   Risk Perception and Implications for Floodway Operation

Although the federal flowage easement has been part of the deeds of the lands within the floodways for decades, the fact that the land is explicitly designated for inundation did not prevent local interests from attempting to stop use of the New Madrid Floodway in 2011, nor has it discouraged the explosion of residential development within the West Atchafalaya Floodway over the past two decades. Residents in the West Atchafalaya Floodway may consider its chances of being used for its designated purpose

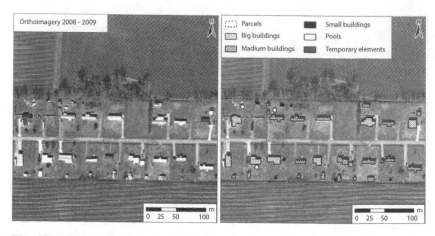

Fig. 2.5 Orthoimage of 2008–2009 of an area of residential development within the West Atchafalaya Floodway (*left*) and same view with various types of encroachments identified (*right*), at coordinates 91°50'9.288"W and 30°59'13.533"N (Source: US Geological Survey; accessed at earthexplorer.usgs.gov, used by permission)

to be low, or may simply be fatalistic about the potential of flooding, a common reaction to flood risk (May and Williams 1986: 5). The West Atchafalaya Floodway now contains houses and swimming pools, besides the original swampland. Such encroachments can interfere with the operation of the floodways (1) by making flood managers reluctant to activate floodways, because of anticipated public resistance, and (2) because of the potential for buildings within the floodway to locally increase hydraulic roughness and reduce conveyance of the floodway. In urban areas, buildings strongly affect flood flow paths (Schubert and Sanders 2012), but the potential effect of buildings on flow resistance in a broad floodway has not (to our knowledge) been analyzed.

2.3.5   Conclusion

While there have been many attempts in the USA at the national level to reduce flood losses through land-use planning, these have not always been supported at the state level and commonly have been circumvented at the local level where land-use decisions are made. Following the disastrous 1993 Upper Mississippi flood, a federal interagency floodplain management review committee (IFMRC 1994) concluded that "The federal government certainly must provide leadership and be a financial supporter of appropriate activities. States, and, as delegated to them by the states, communities, must accept responsibility for land-use planning and should be guiding development in the floodplain" (Galloway 1997: 84–85).

The roles of states in managing floodplains have varied widely across the nation and over time. The NFIP specifies that development be regulated within floodplains inundated by the 100-year flood, but in California, the Central Valley Flood Protection Act of 2008 required a higher standard (the 200-year flood) for urban areas in the Central Valley. NFIP guidelines prohibit new construction in the floodplain that would raise flood elevations by more than 30 cm (1 foot). While some states have legislated stricter standards, in 2004 the state of Missouri took the opposite step and "passed legislation that prohibits any county from setting any threshold stricter than the 1.0-foot limit," thereby contributing to extensive recent floodplain development near St. Louis and elsewhere in the state (Pinter 2005). In contrast to the lack of national-local coordination in the USA, ongoing implementation of the recently adopted Floods Directive in the European Union illustrates a systematic, supra-national approach, which requires systematic mapping of flood risk in all member states and development of measures to reduce risk (Serra-Llobet et al. 2016).

While the basic dilemma of national-local conflicts in land-use management on floodplains is not unique to the examples presented here, conflicts over land use in the New Madrid and West Atchafalaya floodways are particularly compelling (Kondolf and Lopez-Llompart 2018). These are essential components of a river-wide system to manage floods on a large, inter-state river, whose operation reflects national interests. Despite the government's flowage easements, landowners in the New Madrid Floodway objected to inundation of their lands and, through their elected state representatives, attempted to prevent operation of the bypass during the 2011 flood, and have sought to prevent backwater flooding during smaller floods. Despite the government's flowage easements, there has been a threefold increase in residential and commercial development within the West Atchafalaya Floodway since the late 1960s. These new buildings are permitted by local jurisdictions under their land-use authority, but these local decisions have potential to compromise effective operation of a flood control system of national importance.

2.4   Mississippi River Navigation System: A Major Contributor to Flooding

Melissa Samet

2.4.1   Introduction

As described in prior sections of this chapter, the Mississippi River system has been highly altered by agricultural, industrial, and urban land uses, and by extensive modifications for flood control and navigation. As noted above, the modifications to the river system for flood control have been undertaken principally by the US Army Corps of Engineers. In addition to its responsibility to protect communities from flooding, the Corps is charged with restoring portions of the Mississippi and its inland and coastal floodplain, and with reducing impacts to the river and its wetlands when issuing permits for activities of other entities. Notably, the Corps is also responsible for constructing, maintaining, and operating a major navigation channel on the river, extending from the Gulf of Mexico to Minnesota. The navigation responsibility typically conflicts with the Corps' flood damage reduction and restoration responsibilities because the structures built to improve navigation have deleterious impacts on riverine habitats and can increase flood risks by raising water levels during floods.

The Corps' significant level of control and oversight over the river places it in a unique position to advance comprehensive ecological and hydrological restoration of the river and its floodplain to benefit riparian communities and important populations of fish and wildlife. The Corps could make significant progress toward such restoration by developing a fundamentally new approach to operating and maintaining the navigation system.

2.4.2   The Upper Mississippi River Navigation System

The Upper Mississippi River navigation system runs 1394 km (866 miles) from Minneapolis, Minnesota, to the confluence of the Ohio River at Cairo, Illinois (USACE 2017c). The Upper Mississippi River navigation system includes a stretch of river that the Corps refers to as the Middle Mississippi River. The Middle Mississippi runs 314 river km (195 miles) from the confluence of the Missouri River north of St. Louis, Missouri, to the confluence of the Ohio River near Cairo, Illinois (USACE 2017a).

Above St. Louis, the Corps has created a commercially navigable channel through the construction of 29 locks and dams that have transformed the once free-flowing river into a series of highly manipulated pools (Fig. 2.6). There are no locks and dams on the Middle Mississippi River portion of the navigation system. Instead, the Corps has created a commercially navigable channel by heavily constricting the river through hundreds of miles of river training structures (wing dikes, bendway weirs, chevrons) and revetment.

Navigation is maintained through this Upper Mississippi River system by regular dredging of the navigation channel (and disposing of those dredged materials), regulating water flow through the system's locks and dams, constructing new river training structures to force the river into a deeper and narrower channel, placing additional revetment on the river's banks to eliminate natural lateral movement, and operating and maintaining the system's locks and dams.

Construction, maintenance, and operation of the Upper Mississippi River navigation system has fundamentally changed the way the river functions, causing highly significant and well-recognized harm to the environment. A 1999 US Geological Survey report concluded that the Army Corps' navigation management practices have destroyed critical habitats including the rivers' backwaters, side channels, and wetlands; altered water depth; destroyed bathymetric diversity; severely impacted native species; and caused the proliferation of non-native species (USGS 1999). A Biological Opinion issued by the US Fish and Wildlife Service in 2000 determined that key protections were needed to prevent the ongoing management of the river's navigation system from jeopardizing the continued existence of the pallid sturgeon and the Higgins eye pearly mussel (USFWS 2000). The adverse impacts of navigation management were highlighted again in a 2008 US Geological Survey report, which found that the Army Corps' management continued to fundamentally alter the river's hydrologic regime, cause a loss of connectivity to the floodplain, and create high sedimentation rates that had already caused "a substantial loss of habitat diversity" in the system over the past 50 years (USGS 2008).

These adverse impacts are not limited to damage caused by the Upper Mississippi River locks and dams and regular dredging. Significant environmental damage has also been caused by extensive construction of river training structures and revetment in the Middle Mississippi River (USGS 1999, 2008; USFWS 2000).

Fig. 2.6 Mississippi River Lock and dam 27, oblique aerial view from the north (Illinois on the *left*, Missouri on the *right*) (Source: US Army Corps of Engineers, reprinted from Alexander et al. 2012)

2.4.3   River Training Structures in the Middle
Mississippi River

The Corps has constructed more than 1375 wing dikes, bendway weirs, chevrons, and similar structures in the Middle Mississippi River (between river km 290 and 60 (miles 180 and 37)), which amount to 2.4 km (1.5 miles) of river training structures for each mile of this river reach. More than 12,192 m (40,000 feet) of wing dikes and bendway weirs were added in the three years leading up to the great flood of 1993. Many more structures have been added since then, including at least 23 new chevrons between 2003 and 2010.

River training structures are used to reduce the need for, and costs of, navigation dredging by creating a "confined and accelerated flow in the central channel," which causes the channel to incise (downcut), which in turn leads to lower water levels during low flows at most locations (Pinter et al. 2010).

However, at flood flows (flows equal to four or more times the average annual discharge level), these same structures increase water levels by creating "backwater effects upstream of these structures" across a full spectrum of discharges (Pinter et al. 2010). These flood impacts are typically overlooked when evaluating flood risks and flood damage reduction solutions for the Mississippi River, but they should not be, as they pose very real risks to Mississippi River communities.

In the Middle Mississippi, river training structures are responsible for flood height increases of up to 4.5 m (15 feet) in some locations and 3 m (8 feet) and more in broad stretches of the Middle Mississippi where the structures are prevalent (Pinter et al. 2010; Remo et al. 2009). These river training structures contributed to the record crests in 1993, 1995, 2008, 2011, and again in 2015. Dangerously, river training structures and levees have so constricted the Middle Mississippi that it now suffers from the flashy flooding typical of a much smaller river (Criss and Luo 2016).

Analysis of a database of more than 8 million discharge and river stage values and a geospatial database of historical engineering infrastructure (locations, emplacement dates, and physical characteristics of over 15,000 structural features constructed along the study rivers over the past 100–150 years) demonstrates that "the largest and most pervasive contributors to increased flooding on the Mississippi River system were wing dikes and related navigational structures" (Pinter et al. 2008a, 2010). The flood stage impacts of river training structures are cumulative; in the Middle

Mississippi, flood stages have increased by more than 10 cm (4 inches) for each 1000 m (3281 feet) of wing dike built within 32 river km (20 miles) downstream. Progressive levee construction and climate- and/or land-use changes have also contributed to stage increases in the Middle Mississippi, but to a much lesser extent (Pinter et al. 2008a, 2010).

2.4.4   Scientific Consensus Regarding Effects of River Training Structures and the Agency Response

More than 50 peer-reviewed studies support the conclusion that river training structures increase flood stages (e.g., Huthoff et al. 2013; Azinfar and Kells 2007, 2009, 2011; Bormann et al. 2011; Yosseff and de Vriend 2011; Paz et al. 2010; Pinter et al. 2008a, 2010; Theiling and Nestler 2010; Criss 2009; Doyle and Havlick 2009; Pinter 2009; Remo et al. 2009; Jemberie et al. 2008; Ehlmann and Criss 2006; Huang and Ng 2007; O' Donnell and Galat 2007; Remo and Pinter 2007; Yosseff 2005; Ettema and Muste 2004; Wasklewicz et al. 2004; Criss and Schock 2001; Smith and Winkley 1996; Belt 1975 and others). Indeed, a recent theoretical analysis shows that increased flood levels caused by wing-dike construction are "consistent with basic principles of river hydro- and morphodynamics," and that even with extremely conservative parameters used in modeling, "the net effect of wing dikes will be higher flood levels" (Huthoff et al. 2013).

Despite the fundamental hydrological science principle that river training structures increase flood stages, and the extensive empirical evidence that this has occurred on the Middle Mississippi River, the Army Corps' St. Louis District has rejected the scientific consensus of this effect. In contrast, the Corps' St. Paul District recently reached an opposite conclusion more in line with hydrological principles, rejecting a river training structure proposal precisely because that district's modeling showed the structures would produce "significant" and "unacceptable flood stage increases" (USACE 2017b).

Corps leadership and the St. Louis District have rejected numerous requests for a National Academy of Sciences study to guide the agency in its evaluation of this critical public safety issue, requests made by independent scientists in 2008 and subsequently by the St. Louis Post Dispatch editorial board, the conservation community, and thousands of members of the public (e.g., Pinter et al. 2008b; St. Louis Post Dispatch 2010; NWF 2012; ASA 2012; USACE 2017a). A National Academy of Sciences study, which would likely cost less than a single river training structure, could provide vital input for protecting river communities and could help restore the public's confidence in the US Army Corp of Engineers' decision-making.

Instead, the US Army Corp of Engineers' St. Louis District has recently recommended that it continue to build new river training structures in the Middle Mississippi through at least 2034, to further reduce dredging costs (USACE 2017a). This recommendation was based on an environmental impact statement that, once again, explicitly rejected the validity of the science demonstrating the flood stage impacts of river training structures. Not surprisingly, this recommendation is strongly opposed by the public, the conservation community, independent scientists, the US Fish and Wildlife Service, and others (USACE 2017a).

The proposed construction of new river training structures will compound the very real risk of catastrophic flooding that already plagues Mississippi River communities. Dangerously, the next wave of river training structure construction is planned for areas that are already at significant risk. The planned Dogtooth Bend project will be built just downstream from a segment of the Len Small Levee that failed during the 2011 floods (on the Illinois side of the river, 32–64 km (20–40 miles) upstream of the Ohio River confluence). The planned Grand Tower project will be built adjacent to the Big Five Levee System, which has been designated as deficient by Corps inspectors (on the Illinois side of the river near Wolf Lake, 107–118 km upstream of the Ohio River confluence).

The proposed construction would also add to the already extensive losses of fish and wildlife habitat by, among other things, destroying at least another 440 ha (1087 acres) of vital border channel habitat. This would bring the total loss of border channel habitat to 40% in the Middle Mississippi River since 1976 alone, without counting losses prior to this year (USACE 2017a).

2.4.5   Conclusion and Recommendations

The US Army Corps is charged with multiple objectives in its management of the Mississippi River. Notably, the Corps' construction and operation of the navigation system have created striking conflicts with the Corps' flood damage reduction and restoration objectives, and the agency continues to utilize technologies and methodologies that favor navigation at the expense of these other vital interests. The Corps has multiple authorities to protect the public and the environment, which could be more effectively used to advance comprehensive restoration of the river and its floodplain. The critical importance of these issues argues for a halt to new structures, and an independent assessment by the National Academy of Sciences.

To initiate a more balanced and sustainable approach to managing the river and its resources, prioritizing public safety, the Army Corps should:

1. Adopt a moratorium on new river training structures that will remain in effect unless it can be proven that new structures will not increase flood risks for Mississippi River communities.
2. Initiate a National Academy of Sciences study on the role of river training structures on increasing flood heights to inform the Army Corps' decision-making.
3. Conduct a scientifically and legally sound environmental review of the full suite of actions carried out by the Army Corps to maintain navigation on the Upper Mississippi River system and develop and adopt a navigation management plan that will protect people and wildlife. To comply with the Congressionally established National Water Resources Planning Policy (42 USC 1962-3), the measures adopted must protect the environment, including by restoring the river's natural hydrologic and ecosystem functions and by mitigating any harm that cannot be avoided.

   This new plan should (1) abandon the construction of new river training structures, unless it has been demonstrated that they will not increase flood risks; (2) abandon the construction of new revetment that will lock more of the river in place and thereby further harm the river's natural functions; (3) remove and/or modify some of the existing river training structures and revetment to reduce flood risks and restore habitat; (4) restore habitat that has been lost to navigation activities over at least the past four decades; and (5) fully mitigate the adverse impacts of past and future navigation maintenance activities.
4. Advance the wide-scale use of natural infrastructure (healthy rivers, floodplains, and wetlands) as a primary tool for water resources management for the Mississippi River and its floodplain, and throughout the country. Natural infrastructure provides a host of vital benefits, including natural flood protection, clean water, wildlife habitat, and recreational opportunities that are a significant economic driver.

2.5   FLOODPLAINS: MEETING THE NEEDS OF PEOPLE AND NATURE

Todd Strole

Floodplains are among the most fertile and biologically rich lands on earth. Floodplains are a vital component of a healthy river system that supports a diversity of species and a dynamic mosaic of habitats including open water, submersed/emergent aquatic vegetation, wet meadows/prairies, and bottomland hardwood forests. With rich soils, abundant water, and verdant plant growth, river floodplains are tremendously productive. Today, a river's natural floodplain is often separated from the river using levees that protect the land from flooding and provide access to this productivity for agriculture but often protect municipalities and other infrastructure as well. Levees effectively disconnect the river from its floodplain, and while often successful for flood protection, there is an environmental cost. Natural floodplain functions that are reduced dramatically from levee construction include storage and conveyance of floodwaters, natural river hydrology, nutrient cycling, carbon sequestration, sediment management, water filtration/purification, groundwater recharge, habitat for plants and animals, and recreation. The river/floodplain connectivity and the functions it supports have largely been ignored in the past, but we are increasingly aware of the need for functional floodplains in healthy river systems.

People have exploited floodplains and their riches to the detriment of nature, and this use and development in floodplains has often resulted in great loss during floods. Today, we have the understanding, the tools, and the opportunity to change our use of floodplains, reduce such tragedies, and improve the balance between nature and people. We understand that floodplains and rivers in a more natural condition play a critical role in meeting our needs. We can make communities safer, restore critical ecosystem functions, and reduce long-term costs of flood control and disaster relief.

A vision for integrated management of floodplains starts with two premises regarding the needs of nature:

1) We must maintain and restore the key natural processes and functions that sustain floodplain and river systems.
2) We can accurately define the areas, features, and conditions that a floodplain and their associated rivers must have in order to maintain healthy ecological systems.

There are also two premises regarding the needs of people:

1) Floodplains and rivers must provide a significant amount of the goods and services necessary to meet human needs.
2) Understanding flood risks and desirable floodplain functions is not enough to change traditional activities and behavior in the floodplain. Rather, economic incentives, disincentives, and multiple benefit solutions are needed.

As we move toward this vision, there are several examples of research and planning efforts that can guide our work. In the context of flooding, engineers and hydrologists are continually improving the ability to model floods and the impacts that various land-use practices and geomorphic changes will have on a river's ability to store and convey floodwaters. A good example of this is modeling work being conducted on the Missouri River, led by the USGS, where differing land management scenarios have been tested for their impact on flood heights (Bitner 2012). They found that levee setbacks (or removal) and river channel widening can significantly lower flood heights for moderate-sized floods, but this impact is diminished as floods become larger. Using this model, the researchers were able to measure the impact that land-use changes following a record flood in 1993 had on flood heights during the flood of 2007. Work like this demonstrates that we have the ability to design landscape changes that will achieve a target flood-carrying capacity in a river.

So if we can design what we need, then the next step is looking for places to apply these landscape changes. When looking for locations where floodplain restoration can be targeted, there are data sets that are wildly available that can be used to guide floodplain mangers. An example of this is a database for the Upper Mississippi River that was developed in a partnership between the Nature Conservancy and the USACE (Strole 2011). The effort produced a database of information regarding floodplain characteristics that would be useful for screening, planning, mapping, and project identification and included data regarding infrastructure, ownership patterns, and natural resource features.

The restoration of natural floodplain habitat features requires additional information, beyond the hydraulics and hydrology used in the design of flood conveyance. An excellent example of using existing data over large geographies in floodplain management and planning is a technique known as hydrogeomorphic modeling or HGM (Heitmeyer 2008). This method has been used extensively in the Mississippi River Valley, producing detailed maps and information on the historic floodplain vegetation, measuring the changes that have occurred over time, and identifying the options for restoration. This method uses elevations, soils, geomorphic surfaces, historical accounts, and hydrology to produce incredibly detailed maps on historic and current conditions in the floodplain. There are other examples of less intense evaluations such as the Land Capability Potential Index used in the Missouri River floodplain to rapidly assess the current land use and its capacity for other uses (Jacobson et al. 2007).

Broader application of these models, data sets, and assessment methodologies will be needed as we move toward floodplain management that integrates both the needs of people and nature. There are differing approaches and strategies that could move society's understanding, appreciation, and ultimately management of floodplains. However, all would likely include some common elements. For example, demonstration projects that provide research opportunities for measuring floodplain functions including their impact and value in conservation and in flood risk management planning would be required. These projects would provide the foundation of information used to communicate these values to floodplain occupants, stakeholders, and decisions makers. This heightened understanding could lead to a political strategy and ultimately inform a national policy that would improve our management of floodplains through regulation, programs, and market systems that connect the provider of floodplain functions to those receiving the benefit in a real market scheme.

This is a monumental task, and change will be slow and difficult due to the number and complexity of state and federal policies, programs, and regulations. These include but are not limited to executive orders that guide the federal government's role in floodplain development; the NFIP; USDA programs from the "Farm Bill" that provide restoration and insurance programs; numerous projects, authorities, and policies of the US Army Corps of Engineers that exert tremendous influence on the management of floodplains; and state and local management and zoning programs. There are good examples of well-intentioned steps, from legislation to government reports, which have urged us toward wise use of floodplains but have fallen short on implementation. Take, for example, *Sharing the Challenge: Floodplain Management into the 21st Century*, otherwise known as the "Galloway Report," which was an extremely comprehensive report following the 1993 flood in the Upper Mississippi and Missouri River system (Galloway 1994). It provided detailed recommendations that would integrate compatible floodplain uses with appropriate flood protection measures in an effort to reduce flood risk in the future. While some progress has been made, nearly 25 years later, we have failed to reach the vision provided in this report. It remains a much respected document that provides relevant guidance for today regarding many of the issues described above.

Wise use of our nation's floodplains is critical as we continue to face increasing flood losses, a climate that is producing more extreme events and a burgeoning population in need of space to live and land to provide food, fiber, and fuel. However daunting the task may be, the information, technologies, and methodologies are available to guide us. Coupling this with new regulations, programs, and markets that build from our successes and apply lessons learned from our past will be the key to success.

REFERENCES

Alexander, J.S., R.W. Wilson, and W.R. Green. 2012. A Brief History and Summary of the Effects of River Engineering and Dams on the Mississippi River System and Delta: *U.S. Geological Survey Circular 1375*, 43 p.

ASA. 2012. Darcy, Jo-Ellen, Assistant Secretary of the Army. 2013. *Jo-Ellen Darcy Letter to National Wildlife Federation*, March 22.

Azinfar, H., and J.A. Kells. 2007. Backwater Prediction Due to the Blockage Caused by a Single, Submerged Spur Dike in an Open Channel. *Journal of Hydraulic Engineering* 134: 1153–1157.

———. 2009. Flow Resistance Due to a Single Spur Dike in an Open Channel. *Journal of Hydraulic Research* 47: 755–763.

———. 2011. Drag Force and Associated Backwater Effect Due to an Open Channel Spur Dike Field. *Journal of Hydraulic Research* 49: 248–256.

Barker, J. 2014. Environmental Groups and Cairo, Ill., Officials Urge EPA to Veto New Madrid Levee. *St Louis Post-Dispatch*, December 16. http://www.stltoday.com/business/local/environmental-groups-and-cairo-ill-ofcials-urge-epa-to-veto/article_5c61d503-2cd0-5b98-9af5-0b7514f6baea.html.

Barry, J.M. 1997. *Rising Tide: The Great Mississippi Flood of 1927 and How It Changed America*, 523 p. New York: Simon and Schuster.

Belt, C.B. 1975. The 1973 Flood and Man's Constriction of the Mississippi River. *Science* 189: 681–684.

Bitner, C. 2012. *Missouri River Flow Corridor, Boonville, MO to Jefferson City, MO Pilot Study 2008 Modeling Summary & Path Forward.* Presented to the Missouri River Flood Task Force, River and Floodplain Subcommittee.

Bormann, H., N. Pinter, and S. Elfert. 2011. Hydrological Signatures of Flood Trends on German Rivers: Flood Frequencies, Flood Heights, and Specific Stages. *Journal of Hydrology* 404: 50–66.

Camillo, C.A. 2012. *Divine Providence: The 2011 Flood in the Mississippi River and Tributaries Project*, Paper 142. U.S. Army Corps of Engineers, Omaha District.

Criss, R.E. 2009. Increased Flooding of Large and Small Watersheds of the Central USA and the Consequences for Flood Frequency Predictions. In *Finding the Balance Between Floods, Flood Protection, and River Navigation*, ed. R.E. Criss and T.M. Kusky, 16–21. Saint Louis: Saint Louis University, Center for Environmental Sciences.

Criss, R.E., and M. Luo. 2016. River Management and Flooding: The Lesson of December 2015–January 2016, Central USA. *Journal of Earth Science* 27 (1): 117–122. https://doi.org/10.1007/s12583-016-0639-y.

Criss, R.E., and E.L. Shock. 2001. Flood Enhancement Through Flood Control. *Geology* 29: 875–878.

Davis, J., D.C. Otto Jr., D.R. Busse, D. Jody Farhat, D.H. Lee, J. Fredericks, and M.C. Sterling. 2017. *Greater Mississippi Basin 2011 Flood. Post-Flood Operational Performance Assessment.* Washington, DC: US Army Corps of Engineers Headquarters.

Dorothy, O., and P. Nunnally. 2015. The New Madrid Levee: A New Take on an Enduring Conflict. *Open Rivers: Rethinking the Mississippi*, No. 1. http://editions.lib.umn.edu/openrivers/article/new-madrid-an-interview-with-olivia-dorothy/.

Doyle, M.W., and D.G. Havlick. 2009. Infrastructure and the Environment. *Annual Review of Environment and Resources* 34: 349–373.

Ehlmann, B.L., and R.E. Criss. 2006. Enhanced Stage and Stage Variability on the Lower Missouri River Benchmarked by Lewis and Clark. *Geology* 34: 977–980.

Ettema, R., and M. Muste. 2004. Scale Effects in Flume Experiments on Flow Around a Spur Dike in a Flat Bed Channel. *Journal of Hydraulic Engineering* 130 (7): 635–646.

FEMA (US Federal Emergency Management Agency). 1980. *Flood Insurance Study of Town of Simmesport, Louisiana (Avoyelles Parish).* https://msc.fema.gov/portal/availabilitySearch?addcommunity=220025&communityName=SIMMESPORT,TWN%20/%20AVOYELLES%20PARIS#searchresultsanchor. Accessed 25 Oct 2014.

Galloway, G.E. 1994. *Sharing the Challenge: Floodplain Management into the 21st Century*, Report of the Interagency Floodplain Management Review Committee. Report to Congress. Washington, DC: US Government Printing Office.

————. 1995. Learning from the Mississippi Flood of 1993: Impacts, Management Issues, and Areas for Research. In *Proceedings of US-Italy Research Workshop on the Hydrometeorology, Impacts, and Management of Extreme Floods*, Perugia.

————. 1997. River Basin Management in the 21st Century: Blending Development with Economic, Ecologic, and Cultural Sustainability. *Water International* 22 (2): 82–89. https://doi.org/10.1080/02508069708686675.

Heitmeyer, M.E. 2008. *An Evaluation of Ecosystem Restoration Options for the Middle Mississippi River Regional Corridor*, Greenbrier Wetland Services Report 08-02, Advance.

Houck, O.A. 1985. Rising Water: The Federal Flood Insurance Program in Louisiana. *Tulane Law Review* 60 (October): 61–63.

Huang, S.L., and C. Ng. 2007. Hydraulics of a Submerged Weir and Applicability in Navigational Channels: Basin Flow Structures. *International Journal for Numerical Methods in Engineering* 69: 2264–2278.

Huthoff, F., N. Pinter, and J.W.F. Remo. 2013. Theoretical Analysis of Stage Magnification Caused by Wing Dikes, Middle Mississippi River, USA. *Journal of Hydraulic Engineering* 139: 550–556.

IFMRC (Interagency Floodplain Management Review Committee). 1994. *Sharing the Challenge: Floodplain Management into the 21st Century*, Report to the Administration Floodplain Management Task Force, 189 pp. Washington, DC: US Government Printing Office, Superintendent of Documents.

Jacobson, R.B., K.A. Chojnacki, and J.M. Reuter. 2007. *Land Capability Potential Index (LCPI) for the Lower Missouri River Valley*, U.S. Geological Survey Scientific Investigations Report 2007-5256, 19 p.

Jadwin, E. 1928. The Plan for Flood Control of the Mississippi River in Its Alluvial Valley. *Annals of the American Academy of Political and Social Science* 135: 34–44.

Jemberie, A.A., N. Pinter, and J.W.F. Remo. 2008. Hydrologic History of the Mississippi and Lower Missouri Rivers Based Upon a Refined Specific-Gage Approach. *Hydrologic Processes* 22: 7736–4447. https://doi.org/10.1002/hyp.7046.

Kondolf, G.M., and P. Lopez-Llompart. 2018. National-local land-use conflicts in floodways of the Mississippi River system. *AIMS Environmental Science* 5 (1): 47–63. https://doi.org/10.3934/environsci.2018.1.47.

Londoño, A.C., and M.L. Hart. 2013. Landscape Response to the Intentional Use of the Birds Point—New Madrid Floodway on May 3, 2011. *Journal of Hydrology* 489: 135–147.

Lopez-Llompart, P., and G.M. Kondolf. 2016. Encroachments in Floodways of the Mississippi River and Tributaries Project. *Natural Hazards* 81 (1): 513–542. https://doi.org/10.1007/s11069-015-2094-y.

Luke, A., B. Kaplan, J. Neal, J. Lant, B. Sanders, P. Bates, and D. Alsdorf. 2015. Hydraulic Modeling of the 2011 New Madrid Floodway Activation: A Case Study on Floodway Activation Controls. *Natural Hazards* 77 (3): 1863–1887. https://doi.org/10.1007/s11069-015-1680-3.

May, P.J., and W. Williams. 1986. *Disaster Policy Implementation: Managing Programs Under Shared Governance*, 198 pp. New York: Plenum Press.

Morton, L.W., and K.R. Olson. 2013. Birds Point—New Madrid Floodway: Redesign, Reconstruction, and Restoration. *Journal of Soil and Water Conservation* 68 (2): 35A–40A.

Mossa, J. 2016. The Changing Geomorphology of the Atchafalaya River, Louisiana: An Historical Perspective. *Geomorphology* 252: 112–127. https:// doi.org/10.1016/j.geomorph.2015.08.018.

MRC (Mississippi River Commission). 2007a. *The Mississippi River & Tributaries Project: Controlling the Project Flood*, Information Paper. http://www.mvd. usace.army.mil/Portals/52/docs/Controlling%20the%20Project%20 Flood%20info%20paper.pdf. Accessed 25 Oct 2014.

———. 2007b. *The Mississippi River & Tributaries Project: Floodways*, Information Paper. http://www.mvd.usace.army.mil/Portals/52/docs/Floodways%20 info%20paper.pdf. Accessed 25 Oct 2014.

NWF (National Wildlife Federation). 2012. *National Wildlife Federation Letter to Assistant Secretary of the Army for Civil Works, Jo-Ellen Darcy*. Merrifield: NWF (National Wildlife Federation).

O'Donnell, K.T., and D.L. Galat. 2007. River Enhancement in the Upper Mississippi River Basin: Approaches Based on River Uses, Alterations, and Management Agencies. *Restoration Ecology* 15: 538–549.

Olson, K.R., and L.W. Morton. 2012a. The Impacts of 2011 Induced Levee Breaches on Agricultural Lands of Mississippi River Valley. *Journal of Soil and Water Conservation* 67 (1): 5A–10A.

———. 2012b. The Effects of 2011 Ohio and Mississippi River Valley Flooding on Cairo, Illinois, Area. *Journal of Soil and Water Conservation* 67 (2): 42A–46A. https://doi.org/10.2489/jswc.67.2.42A.

Opperman, J.J., G.E. Galloway, J. Fargione, J.F. Mount, B.D. Richter, and S. Secchi. 2009. Sustainable Floodplains Through Large-Scale Reconnection to Rivers. *Science* 326: 1487–1488.

Paz, A.R., J.M. Bravo, D. Allasia, W. Collischonn, and C.E.M. Tucci. 2010. Large-Scale Hydrodynamic Modeling of a Complex River Network and Floodplains. *Journal of Hydrologic Engineering* 15: 152–165.

Pinter, N. 2005. One Step Forward, Two Steps Back on U.S. Floodplains. *Science* 308 (5719): 207–208.

———. 2009. Non-stationary Flood Occurrence on the Upper Mississippi-Lower Missouri River System: Review and Current Status. In *Finding the Balance Between Floods, Flood Protection, and River Navigation*, ed. R.E. Criss and T.M. Kusky, 34–40. Saint Louis: Saint Louis University, Center for Environmental Sciences.

Pinter, N., A.A. Jemberie, J.W.F. Remo, R.A. Heine, and B.S. Ickes. 2008a. Flood Trends and River Engineering on the Mississippi River System. *Geophysical Research Letters* 35: L23404. https://doi.org/10.1029/2008GL035987.

Pinter, N., R. Criss, and T. Kusky. 2008b. *Letter from Nicholas Pinter, Robert Criss, Timothy Kusky to Colonel Lewis F. Setliff III, Commander St. Louis District, Corps of Engineers.*

Pinter, N., A.A. Jemberie, J.W.F. Remo, R.A. Heine, and B.A. Ickes. 2010. Empirical Modeling of Hydrologic Response to River Engineering, Mississippi and Lower P. Missouri Rivers. *River Research and Applications* 26: 546–571.

Remo, J.W.F., and N. Pinter. 2007. The Use of Spatial Systems, Historic Remote Sensing and Retro-Modeling to Assess Man-Made Changes to the Mississippi River System. In *Proceedings of International Association of Mathematical Geology*, ed. P. Zaho et al., 286–288. Geomathematics and GIS Analysis of Resources, Environment and Hazards. Beijing: State Key Laboratory of Geological Processes and Mineral Resources.

Remo, J.W.F., N. Pinter, and R.A. Heine. 2009. The Use of Retro- and Scenario-Modeling to Assess Effects of 100+ Years River Engineering and Land Cover Change on Middle and Lower Mississippi River Flood Stages. *Journal of Hydrology* 376: 403–416.

Schubert, J.E., and B.F. Sanders. 2012. Building Treatments for Urban Flood Inundation Models and Implications for Predictive Skill and Modeling Efficiency. *Advances in Water Research* 41: 49–64.

Serra-Llobet, A., E. Conrad, and K. Schaefer. 2016. Governing for Integrated Water and Flood Risk Management: Comparing Top-Down and Bottom-Up Approaches in Spain and California. *Water* 8: 445. https://doi.org/10.3390/w8100445.

Shadie, C.E., and B.A. Kleiss. 2012. *The 2011 Mississippi River Flood and How the Mississippi River & Tributaries Project System Provides Room for the River.* 2012 EWRI-ASCE World Environmental and Water Resources Congress, Albuquerque.

Smith, L.M., and B.R. Winkley. 1996. The Response of Lower Mississippi River to River Engineering. *Engineering Geology* 45: 433–455.

St. Louis Post Dispatch. 2010. Editorial Board. *Corps of Engineers River Project Needs Independent Review*, August 27.

Strole, T.A. 2011. *Floodplain Restoration Planning Under NESP, Internal Report.* St. Louis: The US Army Corps of Engineers, St Louis District.

Taylor, B. 2007. Construction Ordered Stopped on New Madrid Floodway. *Southeast Missourian*, September 18. Associated Press.

Theiling, C.H., and J.M. Nestler. 2010. River Stage Response to Alteration of Upper Mississippi River Channels, Floodplains, and Watersheds. *Hydrobiologia* 640: 17–47.

USACE (US Army Corps of Engineers). 1938. *West Atchafalaya Floodway*, Typescript Report by the Second New Orleans District. New Orleans: US Army Corps of Engineers.

————. 2015. *St. Johns Bayou and New Madrid Floodway Project*. http://www.mvm. usace.army.mil/Missions/Projects/StJohnsBayouandNewMadridFloodwayProject. aspx. Accessed 4 May 2015.

————. 2017a. U.S. Army Corps of Engineers St. Louis District, *Final Supplement I to the Final Environmental Statement Mississippi River Between the Ohio and Missouri Rivers (Regulating Works) (May 2017) and Appendices*, Available at http://mvs-wc.mvs.usace.army.mil/arec/Documents/SEIS/Final_SEIS/ Regulating_Works_Final_SEIS_Main_Report_and_App_A_through_H.pdf. Visited 4 Aug 2017.

————. 2017b. U.S. Army Corps of Engineers St. Paul District, *DRAFT Letter Report and Integrated Environmental Assessment, Lower Pool 2 Channel Management Study: Boulanger Bend to Lock and Dam No. 2* (DRAFT June 2017) at 42. Available at http://www.mvp.usace.army.mil/Home/PN/ Article/1219079/draft-lower-pool-2-channel-management-study/. Visited 22 June 2017.

————. 2017c. *Upper Mississippi River Locks & Dams* (2017). Available at http:// www.mvr.usace.army.mil/Portals/48/docs/CC/FactSheets/MISS/ UMR%20Locks%20and%20Dams%20-%202017%20(MVD).pdf?ver= 2017-05-11-111653-327. Visited 2 Aug 2017.

USFWS (US Fish and Wildlife Service). 2000. *Biological Opinion for the Operation and Maintenance of the 9-Foot Navigation Channel on the Upper Mississippi River System*. Rock Island: USFWS (US Fish and Wildlife Service).

USGS (US Geological Survey). 1999. *Ecological Status and Trends of the Upper Mississippi River System 1998: A Report of the Long Term Resource Monitoring Program*. La Crosse: USGS (US Geological Survey).

————. 2008. In *Status and Trends of Selected Resources of the Upper Mississippi River System, Technical Report LTRMP 2008-T002. 102 pp + Appendixes A–B*, ed. B.L. Johnson and K.H. Hagerty. La Crosse: Upper Midwest Environmental Sciences Center.

Wasklewicz, T.A., J. Grubaugh, and S. Franklin. 2004. 20th Century Stage Trends Along the Mississippi River. *Physical Geography* 25: 208–224.

Wittenberg, A. 2017. Exiting Obama Hamstrings Hot-Button Flood Project. *E&E News*.

Yossef, M.F.M. 2005. *Morphodynamics of Rivers with Groynes*. Delft: Delft University Press.

Yossef, M.F.M., and H.J. De Vriend. 2011. Flow Details Near River Groynes: Experimental Investigation. *Journal of Hydraulic Engineering* 137: 504–516.

CHAPTER 3

Managing Floods in Large River Basins in the USA: The Sacramento River

Rod Mayer, Timothy Washburn, John Cain,
and Anna Serra-Llobet

Abstract The Sacramento River was an important area for flood management, with involvement by the federal government back to the nineteenth century, starting with the same levees-only approach as initially used on the Mississippi, but evolving to the current system of flood bypasses implemented in the early twentieth century. After the 1986 and 1997 floods (and a 2003 court decision holding the state liable for damages from levee failures), the California legislature enacted a set of reforms in 2007 that included an enhanced flood protection standard for urban areas of the

R. Mayer
HDR, Inc., Omaha, NE, USA

California Department of Water Resources, Sacramento, CA, USA

T. Washburn
Sacramento Area Flood Control Agency, Sacramento, CA, USA

J. Cain
American Rivers, Berkeley, CA, USA

A. Serra-Llobet (✉)
University of California Berkeley, Berkeley, CA, USA

Aix-Marseille University, Marseille, France

© The Author(s) 2018
A. Serra-Llobet et al. (eds.), *Managing Flood Risk*,
https://doi.org/10.1007/978-3-319-71673-2_3

43

Central Valley; maps showing 100- and 200-year floodplains; and programs to set back some levees and strengthen others. The National Flood Insurance Program approach is a poor fit for agricultural areas, and the state is exploring ways to modify its application to agricultural areas. Flood bypasses, crucially important infrastructure that also provide valued wildlife habitat, will be expanded to yield multiple benefits.

Keywords Sacramento River • Central Valley • Flood bypasses • Levee safety

3.1   INTRODUCTION

Rod Mayer, Timothy Washburn and Anna Serra-Llobet

3.1.1   *The Sacramento River Basin*

The Sacramento River is the largest river in California, by length and discharge, draining about 68,635 km^2 (26,500 mi^2). The river flows south from the Klamath Mountains for 595 km (370 miles) to the Sacramento-San Joaquin River Delta and San Francisco Bay (Fig. 3.1). For most of its length, it flows within the fertile agricultural region bounded by the Coast Ranges and Sierra Nevada known as the Sacramento Valley. Before the development of the Sacramento Valley, during winter and spring of wet years, six natural overflow basins of the river would become inundated, creating an "inland sea" over 240 km (150 mi) long and 64 km (40 mi) wide (Kelley 1998).

The natural runoff of the Sacramento River catchment is 27 km^3 (22 million acre feet) per year, an average flow rate of about 850 m^3/s (30,000 ft^3/s). Despite the attenuating impact of several large dams in the catchment, flows in the river can reach up to 18,000 m^3/s (650,000 ft^3/s) during the rainy season, equal to the flow of the Mississippi River (US Department of Interior/US Geological Survey 2011).

Discovery of gold in the American River, a tributary of the Sacramento River, started the California Gold Rush in 1848. Hydraulic mining for gold between 1853 and 1884 in the Sierra Nevada and its foothills removed billions of cubic meters of sediment, much of which accumulated in downstream creeks and rivers, including the Sacramento River and many of its tributaries, greatly exacerbating natural flooding (Lund 2012).

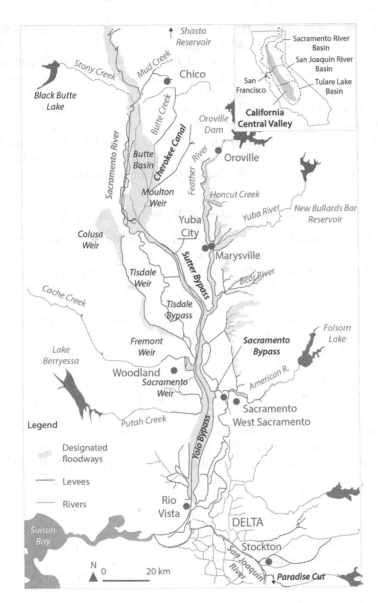

Fig. 3.1 Features in the Sacramento River flood system (Source: Redrawn from US Army Corps of Engineers and California Department of Water Resources; accessed at spk.usace.army.mil/Missions/Civil-Works/Sacramento-River-GRR)

Since the 1940s the Sacramento River catchment has been intensely developed for water supply and the generation of hydroelectric power. Today, large dams impound the river and almost all of its major tributaries. The Sacramento River provides agricultural, industrial, and municipal water throughout the Sacramento Valley and is exported to the San Joaquin Valley, the San Francisco Bay Area, and metropolitan Southern California through the pumps and canals of giant state and federal water projects. Along with its tributaries, the Sacramento River provides water to over half of California's population. These changes to the Sacramento River and its tributaries from their natural state have caused the decline of its once-abundant fisheries.

3.1.2   The Sacramento Valley: A Highly Engineered System

The Sacramento and San Joaquin Valleys, together known as the Central Valley, contain an extensive system of levees built over the course of more than a century to protect nearly 809,000 ha (2 million acres) from flooding caused by an inland sea that historically formed in many winters. Construction of this levee system was commenced by local interests in the latter half of the nineteenth century. However, the scale and complexity of this undertaking required a degree of central planning and administration that could only be provided by the State of California in partnership with the federal government. Thus, by the middle of the twentieth century, what had begun as a disjointed assemblage of local levees became the State Plan of Flood Control (SPFC) encompassing 2570 km (1600 mi) of state-federal (hereafter, federal) levees the vast majority of which are in the Sacramento Valley. Together with an extensive network of non-federal levees that have retained their local character, SPFC protects about 1 million people and nearly $70 billion of infrastructure in the Central Valley (CDWR 2012a).

Above the valley floor, large dams and reservoirs constructed on the Sacramento River and most of its tributaries provide water storage, flood protection, power, and recreation. Dedicated flood storage space during fall through spring at these reservoirs enhances the flood protection provided by the downstream levees. Reservoir operation rules for fall through spring were developed by the US Army Corps of Engineers (USACE), who provided part of the funding for the construction of the dams.

3.1.3   Flood Risk Management in California: Governance System

Construction of the levees comprising the SPFC was largely completed by the USACE in the 1950s and 1960s at which point these levees were turned over to the California Reclamation Board (Board—now the Central Valley Flood Protection Board) for operation and maintenance. The Board then entered into a series of agreements with local levee districts, reclamation districts, and the Department of Water Resources (DWR) to perform the required operation and maintenance. Under this arrangement, DWR is responsible for ensuring that federally designated channel capacities in the Sacramento Valley are maintained for the conveyance of design flood flows. DWR is also directly responsible for maintaining about 482 km (300 mi) of federal levees in the Central Valley. The remaining 2092 km (1300 mi) of the federal levee systems in the valley are operated and maintained by local reclamation districts and levee maintenance districts.

In addition to the federal levee systems, the Board regulates approximately 1931 km (1200 mi) of floodways along the Sacramento and San Joaquin Rivers and their tributaries. The purpose of the designated floodways is to preserve historic flooding patterns and prevent encroachments that would redirect or raise flood waters onto other properties and structures.

There is no state responsibility for levee or channel operation and maintenance outside the Central Valley. However, DWR provides statewide planning assistance for flood risk management and manages state subvention funding that reimburses local agencies for most of the non-federal cost share on federal flood control project construction. DWR is also responsible for assisting in flood emergencies statewide.

In this chapter, we review the flood management system in the Central Valley of California, recent large floods and responses to them, and the Central Valley Flood Protection Plan (CVFPP), which increased levee standards for urban areas in the valley, required stronger building codes, and improved floodplain evaluation and delineation. We also consider changes needed in the application of the National Flood Insurance Program (NFIP) to maintain the viability of agriculture. John Cain reviews the flood risk management, ecological, and social benefits of flood bypasses in the Central Valley and proposals for the expansion of the bypass system.

3.2   MANAGING FLOODS IN THE CALIFORNIA CENTRAL VALLEY

Rod Mayer and Timothy Washburn

3.2.1   State Plan of Flood Control

Levee construction in the Sacramento Valley began in the 1860s shortly after settlers arrived in the valley and found themselves subject to winter flooding. These early levees were constructed to a variety of engineering standards by local forces with available local funding. Later in the early twentieth century the USACE incorporated these local levees into a single system with a standard grade and dimension for passage of certain design flows based on observations from the 1907 and 1909 floods in the Sacramento Valley. This effort was facilitated by the Board, which provided USACE with the lands, easements, and right of way required for levee construction. The resulting SPFC greatly improved the reliability of the levee system. However, the SPFC did not fully remediate the poor quality of the original levees, many of which were built on poor foundations subject to heavy under-seepage.

SPFC levees were set close to the river channel in order to improve navigation by having the rivers scour hydraulic mining sediments. The design of the system assumed no levee failures, but included five engineered diversions and one natural overflow diversion. The natural diversion is to Butte Basin at the upper end of the levee system. The five engineered diversions include two additional diversions to Butte Basin (Moulton and Colusa Weirs), one diversion to the Sutter Bypass (Tisdale Weir), and two diversions to the Yolo Bypass (Fremont and Sacramento Weirs) (Fig. 3.1). All of the engineered diversions included the acquisition of property rights to support the diversions. The deliberate planning, construction, and maintenance of the diversions ensured that they would function during flood conditions and serve as reliable features of the flood project.

3.2.2   Standard Levee Design

Initially, the river channel and bypass levees in each segment of the system were constructed based on a standard geometry. The levees were designed with a predetermined freeboard allowance tied to specified flows and associated water surface elevations, generally matched to the 1907 and 1909 floods, adjusted for loss of natural floodplain storage by construction of the levee system. Over time, the standard levee section was increased because of numerous levee failures. The minimum standard levee changed from a levee with a top width of 3 m (10 feet) to one with a top width of 6 m (20 feet). In addition, the design flows were modified substantially on the Feather and American rivers. This was the result of floods that occurred after 1909, which demonstrated these rivers could produce substantially greater flows than occurred during the 1907 and 1909 floods.

Because numerous levee failures occurred along the Feather River levees between 1920 and 1934, these levees were set back and enlarged to accommodate greater flows. These changes were summarized in memorandums issued by the USACE, which define the minimum freeboard requirements for each segment of the SPFC, collectively referred to as the "USACE 1957 Profile."

3.2.3   Heightened Protection for Urban Areas

Over the years, the capacity of the SPFC was greatly expanded by the construction of five major multiple-purpose reservoirs (Shasta, Black Butte, Oroville, New Bullards Bar, and Folsom) containing 3.3 billion m³ (2.7 million acre feet) of flood control storage space. Institutional support for these reservoirs reflected the commonly shared view during the 1950s and 1960s that concentrated urban populations should be afforded a very high level of flood protection so as to avoid catastrophic losses of life and property. Urban development in the Sacramento Valley generally occurred by the expansion of the historic gold rush settlements of Sacramento, West Sacramento, Yuba City, and Marysville. Protecting Sacramento and West Sacramento was central to the design and construction of Folsom Dam on the American River, while protecting Yuba City and Marysville following the devastating flood of 1955 galvanized construction of New Bullard's Bar Dam on the North Fork of the Yuba River and Oroville Dam on the Feather River.

The flood control operations at these facilities were generally designed to provide these urban areas with "standard project flood protection"— defined as protection from the most extreme flood event that could be considered reasonably foreseeable given the meteorological and hydrological character of the surrounding catchment. Agricultural areas in Sacramento Valley benefited from these operations, but the concept of standard project flood protection had a specifically urban orientation and represented an early expression of the urban/rural dichotomy that has become central to the SPFC (Senate Bill 5, 2006–07).

3.2.4   National Flood Insurance Program

The NFIP was created in the early 1970s to address a growing national concern as to how to manage urban development in areas subject to flooding. After much debate, Congress concluded that it would not be practical or economically feasible to tie the national flood insurance pool and flood-plain development requirements in general to the concept of standard project flood protection. Instead, the 100-year flood was chosen as the standard for administering the NFIP. As cities and counties in the Sacramento Valley sought to enter the program in the late 1970s, the Federal Emergency Management Agency (FEMA) looked to USACE to characterize the protective capacity of the flood control system in the various basins and sub-basins comprising the SPFC. Based on the historic design of the system including the augmentation provided by the multi-purpose reservoirs, USACE concluded that the 100-year flood was generally contained within the USACE 1957 Profile, and thus there was no reason to believe that such a flood could not be safely contained wherever SPFC levees comprised the line of defense. This conclusion allowed virtually all of the lands within the SPFC to enter the NFIP with flood insurance rate maps (FIRMs) indicating areas protected by levees with a moderate risk of flooding (less than 1 percent annual risk of flooding).

3.2.5   The Flood of 1986

The record flood of 1986 severely tested USACE's hypothesis. Although this flood was significantly larger than the 1907 and 1909 floods, as anticipated the availability of reservoir storage largely prevented flows in the system from exceeding the design of the SPFC. Nevertheless, numerous project levees experienced unexpectedly severe stress and some failed. A notable levee failure occurred along the south levee of the Yuba River, subjecting the communities of Linda and Olivehurst to deep flooding. This experience caused USACE, the state, and their local partners to perform a series of levee evaluations on the SPFC levees and to implement system-wide improvements aimed at addressing identified vulnerabilities particularly levee through-seepage. These improvements were generally implemented under USACE's authority to remedy design deficiencies in federal project levees, where benefits exceeded costs.

The 1986 flood also triggered a series of engineering feasibility studies focusing on opportunities for additional flood risk reduction in the urban areas of the Sacramento Valley—Sacramento (including Natomas), West Sacramento, Marysville, and Yuba City. These studies proceeded under the reforms enacted as part of the Water Resources Development Act of 1986, which required more substantial non-federal cost sharing and tied the federal interest to maximizing net economic benefits. The studies produced urban levee improvement projects particularly in Sacramento/ Natomas and West Sacramento that enabled these areas to address levee height deficiencies identified in the aftermath of the 1986 flood.

3.2.6   The Flood of 1997

The flood of 1997 essentially equaled the runoff and resulting water surface elevations produced by the 1986 flood in the levee confined channels comprising the SPFC. The improvements carried out as part of the response to the 1986 flood helped to maintain the stability of most segments of the system, but as in 1986 several levee systems experienced severe stress and some failed. This time the post-flood assessment pointed to a risk factor not historically addressed by the design of the SPFC levee system—levee under-seepage. Prior to 1997, flood managers regarded under-seepage as a risk that could be adequately addressed through levee monitoring and flood fighting. In the aftermath of the 1997 flood, USACE and its state and local partners determined that SPFC levees should be designed to address the risk of under-seepage where geotechnical data indicated potential vulnerability. This determination more than any other in the post-1986 era has contributed to the emergence of distinctly urban and rural levee systems since it is prohibitively expensive for non-urban levee districts to carry out the geotechnical investigations necessary to identify under-seepage vulnerabilities or to implement the improvements (cut-off walls, seepage berms, or relief well systems) necessary to address identified problems (CDWR 2010a).

3.2.7   2003 Paterno Decision

In 2003 the State Appellate Court ruled against the state in the *Paterno* Decision, resulting in a $500 million judgment against the state for the 1986 levee failure on the Yuba River. In response, the California Legislature asked DWR to develop a report on the state's potential liability for the SPFC levee system. In early 2005, DWR released *Flood Warnings: Responding to California's Flood Crisis* (CDWR 2005). This report documented several key challenges:

• The levees were built according to outdated practices
• The levees are old and deteriorating

- There is increasing development in the floodplains behind these levees
- Court decisions have increased the state's liability should a levee fail
- State, local, and federal funding for flood management is inadequate and has declined

In response to this report and to the flooding of New Orleans in 2005, the California State Legislature passed, and voters approved nearly $5 billion of general obligation bond funds in 2006 for flood management investments.

3.3   THE CENTRAL VALLEY FLOOD PROTECTION PLAN

Rod Mayer Timothy Washburn

In 2007, the legislature enacted a package of flood reform bills that directed DWR to perform numerous flood risk management activities, including a thorough evaluation of the SPFC levees, floodplain mapping, and a revitalization of the SPFC through development of a CVFPP. These reforms are set forth in the Central Valley Flood Protection Act of 2008 (Senate Bill 5, 2006–07). The act includes the following findings:

- The Central Valley of California is experiencing unprecedented development, resulting in the conversion of historically agricultural lands and communities to densely populated residential and urban centers.
- The legislature recognizes that by their nature, levees, which are earthen embankments, typically founded on fluvial deposits, cannot offer complete protection from flooding but can decrease the frequency of flooding.
- The legislature recognizes that the level of flood protection afforded to rural and agricultural lands by the original flood control system would not be adequate to protect those lands if they are developed for urban uses and that a dichotomous system of flood protection for urban and rural lands has developed through many years of practice.
- The legislature further recognizes that levees built to reclaim and protect agricultural land may be inadequate to protect urban development unless those levees are significantly improved.

- Cities and counties rely upon federal floodplain information when approving developments, but the information available is often out of date and the flood risk may be greater than that indicated using available federal information.
- The legislature recognizes that the current federal (FEMA) flood standard is not sufficient for urban and urbanizing areas within flood-prone areas throughout the Central Valley.

Based on these findings, the act establishes a new flood protection standard for urban areas (defined as "developed areas in which there are 10,000 residents or more") located in levee-protected floodplains in the Central Valley. This new "urban level of flood protection" is defined as "the level of protection that is necessary to withstand flooding that has a 1-in-200 chance of occurring in any given year using criteria consistent with, or developed by, the Department of Water Resources."

This approach represents a reversal of the outcome of the NFIP debate of the 1970s on the flood protection standard necessary to support urban development—at least in the levee-protected floodplains of the Central Valley. Although not explicitly referred to as standard project flood protection, the state's new "200-year" urban standard, particularly as it is being developed by DWR, is in many ways the equivalent of the SPFC standard, which Congress rejected when the NFIP was created.

3.3.1   Central Valley Flood Protection Plan

The Central Valley Flood Protection Plan (CVFPP) is the vehicle by which the policies adopted by the legislature are being implemented. It is estimated that the rural (non-urban) areas in the Central Valley occupy the majority of the lands in the plan area and contain as many as 50,000 residents occupying a range of rural landscapes including more than two dozen "small communities." The CVFPP makes it clear that the cost of improving these levees to current engineering standards (including resistance to under-seepage) would be prohibitively expensive. Thus, the plan calls for a more limited program of rural levee improvements focused on levee accessibility for flood monitoring and flood-fighting activities, erosion protection, and remediation of extraordinary embankment and foundation stability problems. The CVFPP also anticipates that where feasible rural small communities will be provided with a combination of structural and non-structural improvements sufficient to meet the minimum requirements of the NFIP.

These improvements could include compartment levee systems, structure raising, and other forms of flood proofing (CDWR 2012b).

Urban and urbanizing areas on the other hand are to be protected through a combination of improvements to existing local levee systems and a series of system improvements including physical and operational improvements to existing dams and reservoirs, expansion of existing bypass channels, and creation of new bypass channels. Urban levees are to be designed to safely contain flows up to the 200-year water surface elevation with 1 m (3 feet) of freeboard. This design water surface elevation is to be calculated without adjustment for the possibility of upstream levee failures (CDWR 2012b).

An important concept presented in DWR's criteria is that, in leveed areas, it may be possible to find that part of a levee system can effectively protect an area of land where the land-use decision is under consideration. This would only work in relatively long leveed areas where the land-use decision is toward the upper end of the leveed area. The ability to make such a finding for only part of a levee system not only reduces the cost of achieving an urban level of flood protection for the urban or urbanizing portion of the leveed area but also offers the ability to preserve agricultural land use in the lower portion of the leveed area. This concept is different than current USACE and FEMA procedures regarding levee certification and accreditation for 100-year flood protection, which require the entire levee system to be included.

3.3.2   Handbook for Local Communities

One of the principal objectives of the 2007 reform package is to establish stronger connections between local land-use decisions and flood risk. Toward this end, in 2010, DWR released *A Handbook for Local Communities*, which helps cities and counties throughout California comply with these new laws (CDWR 2010b). The handbook details the responsibilities of local jurisdictions, timelines for compliance, and "crosswalks" for certain requirements (available at: www.water.ca.gov/localfloodriskplanning).

3.3.3   Building Code

In addition, DWR has submitted building code updates to the California Building Standards Commission (CBSC) for new construction in deep (greater than 1 m (3 feet)) 200-year floodplains behind SPFC facilities. To our knowledge, California is the only state proposing building code standards applicable outside of Special Flood Hazard Areas (SFHAs). DWR intends to make proposals over multiple code update cycles. In 2010, DWR submitted its first proposal, which provides for new single family residential structures to have a safe "flood evacuation location" and a clear path to it, so as to avoid the situation where people drown in their homes and attics—such as occurred in New Orleans in 2005. Because 200-year floodplain maps are not yet available, the CBSC adopted this code provision in the voluntary section of the code. After 200-year floodplain maps are available, DWR plans to request that this code provision be made mandatory.

3.3.4   Flood Risk Notification

DWR has also initiated a flood risk notification program that is unique to California. Under this program, there is an annual notice mailed to all property owners in areas protected by SPFC facilities. The notice provides information about flood risk, nearby streams and levees, and actions to take in preparing for a flood—including purchase of flood insurance. An accompanying website provides additional information, including whether deep (>1 m (3 feet)) or shallow flooding may occur at any particular address.

3.3.5   Central Valley Floodplain Evaluation and Delineation

In support of this flood risk notification effort, DWR has initiated a floodplain evaluation and delineation program to develop new 100-year, 200-year, and 500-year floodplain maps for areas associated with the SPFC. Floodplain maps are one of the most important tools for informing land-use decision-makers, floodplain residents, and the general public about areas subject to flooding. Unfortunately, most federal levees in Sacramento Valley were "grandfathered" on floodplain maps developed decades ago by FEMA, showing the areas behind these federal levees as being adequately protected from the 100-year flood. These maps are woefully out of date and do not reflect current knowledge of the condition of these levees.

The components of DWR's program are (1) improved topography through new LiDAR and ground surveys, (2) hydraulic evaluation through new riverine and overland hydraulic models (under development), (3) updated floodplain delineations based on existing hydrology and new geotechnical data to identify levee breaches for floodplain modeling, with updated hydrology under development for future use.

3.3.6   *Levee Evaluations*

Finally, DWR has prepared a detailed report evaluating the levees comprising the SPFC. This Flood Control System Status Report was completed in 2011 based on work performed by DWR between 2006 and 2011 (CDWR 2011). The report concluded that half of urban levees do not meet current engineering criteria, 60 percent of non-urban levees have high potential for failure, and half of evaluated channels cannot pass design flows (Fig. 3.2).

3.3.7   *Conclusion*

An essential premise of the 2007 reforms is that it is not affordable or economically justified to build our way out of flood risk. Although we will invest heavily in our flood protection systems, especially in our urban levee systems, we also have to work to reduce consequences when floods do occur and to limit future intensification of development in floodplains. Developing and providing good flood risk information to decision-makers and the public is critical in helping to manage the consequences. We also need to limit development in the non-urban floodplains, making good investment choices and finding new ways to help sustain agricultural econ-

Fig. 3.2   Jones Tract levee failure in June 2004, California (Courtesy of California Department of Water Resources)

omies. As we make our investments, we need to employ concepts of integrated water management to develop projects that provide multiple benefits wherever feasible.

3.4 ADAPTING FLOOD INSURANCE FOR AGRICULTURAL AREAS

Timothy Washburn Rod Mayer

3.4.1 NFIP Policy and Agricultural Areas in the Central Valley

The Special Flood Hazard Area (SFHA), the land area covered by the floodwaters of the 100-year flood, is the area where the NFIP's floodplain management regulations must be enforced and the area where the mandatory purchase of flood insurance applies (FEMA 2017). The Agricultural Floodplain Ordinance Task Force (Task Force) noted that agriculture is one of the most appropriate land uses within an SFHA—but sustainability of agriculture in the Sacramento Valley is being impacted by mapping most rural areas as SFHAs on new FIRMs for the first time.

When an agricultural area is mapped into an SFHA, it results in (1) elevation or floodproofing requirements for new and substantially improved structures, and (2) a requirement to purchase flood insurance policies for structures with federally backed mortgages.

The financial burden can be far greater than the risk exposure as a result of the FEMA practices and policies. The Task Force stated:

- "Insurance premiums are based on the assumption that a non-accredited levee provides no flood protection, when in fact most non-accredited levees provide a substantial amount of flood protection that can be quantified and recognized. Since agricultural areas can rarely afford to have accredited levees, the effect is that many leveed agricultural areas pay insurance premiums that are much higher than the associated flooding risk.
- Insurance premiums for agricultural structures are generally the same as for retail business and industrial structures, which are thought to be more vulnerable to flood damage than agricultural structures.

- Fully wet floodproofed structures are required to pay insurance premiums as if they had no floodproofing.
- Each structure on a parcel is required to have an individual policy with a $250 annual surcharge. Farms typically have far more structures than other types of businesses.
- Low value detached structures associated with agriculture are required to have flood insurance coverage when similar structures associated with a residence would not.
- Insurance premiums for structures in areas protected by reaches of levee that meet all federal requirements are charged at the Zone D rate (referring to areas where there are possible but undetermined flood hazards, as no analysis of flood hazards has been conducted) instead of the lower Zone X (Shaded) rate (which represents moderate- and low-risk areas), if the levee reach happens to be part of a larger levee system.
- Insurance premiums for structures in areas protected by well-studied sound reaches of non-accredited levee are charged at the Zone D rate, the same as areas of undetermined flood risk."

These requirements are inconsistent with the 2012 CVFPP's recognition that through many years of practice, the urban and rural/agricultural levee systems in the Central Valley have evolved into distinctly different systems that require distinctly different levee and floodplain management policies.

3.4.2   Need for NFIP Policy Adjustments in Agricultural Areas

The NFIP lacks policies which specifically address insurance and land-use management requirements in rural agricultural areas. Nevertheless, there are many reasons for developing such policies. The lightly populated levee districts that have been created to foster agricultural production in the Central Valley provide an array of economic and ecological goods and services that warrant public support. First, these lands tend to be very productive. They serve as the core of the rural economies that surround them and they contribute substantially to regional, state, and national economic development. Second, they provide significant habitat value for important wildlife populations particularly migratory birds that have adapted to the agricultural landscape. Third, they provide abundant open space and recreational values. Finally, and perhaps most importantly for this discussion, they provide a significant buffer against the risk of catastrophic flooding in the urban areas of the Central Valley.

For these reasons, it would make sense to develop NFIP policies which contribute to the social and economic sustainability of these agricultural levee districts.

This could be accomplished in a variety of ways including the development of a special agricultural zone designation with accompanying insurance and land management requirements, use of FEMA's Community Rating System (CRS) to complement the special agricultural zone designation, and use of federal credits earned in connection with urban levee improvement projects to mitigate resulting financial impacts on the NFIP.

3.4.3   Agricultural Zone Designation

The concept of developing an agricultural zone designation within the NFIP would help to relieve levee-protected agricultural areas of the most onerous consequences of not being able to meet the modern engineering standards by which urban levee systems have come to be measured. These burdens include (1) the cost of insuring farm residences and other farm structures in a post-Katrina environment in which insurance rates may be subject to sharp increases and (2) restrictions on what can be done with farm residences and structures that need to be replaced or improved. A new agricultural zone designation could provide relief from these burdens in the form of modified insurance rate schedules and structure replacement and improvement requirements appropriately tailored to agricultural conditions.

3.4.4   Rural Community Rating System

In order to complement such an agricultural zone designation, FEMA could consider appropriate modifications to its CRS program to make this program a better fit for the agricultural levee districts in the Central Valley. Here it might be possible to take advantage of the state's status as a participating CRS community. Perhaps for purposes of rating the levee districts under the jurisdiction of the CVFPP, FEMA could consider the flood risk reduction accomplishments of the plan as a whole, rather than looking at each levee district in isolation. The ratings could also account for the CVFPP's small communities program which aims to protect or otherwise flood proof the rural communities that occupy many of the rural levee districts. The residential structures comprising these communities account for a significant percentage of all residential structures in the affected levee districts, and the CVFPP's program of bringing these structures into compliance with NFIP requirements could play an important role in a well-conceived rural CRS program.

3.4.5   Federal Credit Exchange

The state is currently accumulating a substantial amount of federal credit through USACE for initiating urban levee improvement projects around the Sacramento Valley in advance of federal authorization of these projects. In theory, these credits could be used to offset future state contributions to completing these projects and other levee projects in the valley. However, constraints on federal funding for USACE projects are undermining the state's prospects for deploying the credits.

An optional use for these credits would be to apply them to reduce insurance rates in the agricultural areas within the jurisdiction of the CVFPP. Under this arrangement, USACE would affirm the amount of the credits and FEMA would be authorized to apply the credits (perhaps through the CRS program) to reduce insurance rates in the specified agricultural areas. This repurposing of the credits would provide a significant incentive for the state to accelerate improvements to urban levee systems, thereby reducing the risk of widespread property damage. Meanwhile, the financial implications of providing insurance relief to property owners in the agricultural areas could be addressed by allowing FEMA to use the exchanged credits to reduce their interest payments on the NFIP's outstanding debt to the Treasury. This could be considered an appropriate final disposition of the credits since they represent a kind of federal obligation similar to the obligation created by the Treasury's loan to the NFIP. The credits were created by non-federal advances to the federal government made under exigent circumstances to reduce the risk of catastrophic flood damage in urban areas before USACE could position itself to address this risk. Similarly, the NFIP debt to the Treasury was produced by an advance to the NFIP also under exigent circumstances to facilitate recovery from a catastrophic flood before the NFIP could position itself to fully address the needs of the recovery process. It would seem reasonable therefore that the credits could be used to retire a portion of the debt. The resulting reduction in interest payments on the debt would roughly equal the reduction in premium payments that would result from the credit exchange.

3.4.6 Recommendations of the Agricultural Floodplain Ordinance Task Force

In 2015 DWR agreed to fund the work of a Task Force aimed at developing recommendations to FEMA that could be implemented administratively, without changing law or regulation, for improving sustainability of agriculture in leveed SFHAs. The Task Force is comprised of a broad mix of state, local, and non-governmental participants. The following is taken from their report.

"Between February and December 2016 the Task Force developed nine recommendations to FEMA that can be implemented administratively. However, not all member organizations support all of the recommendations. The recommendations are presented in *Recommended Administrative Refinements of the National Flood Insurance Program to Sustain Agriculture as a Wise Use of the Floodplain in Leveed Special Flood Hazard Areas* (HDR 2016).

The following nine recommendations address how FEMA's rules and practices could be modified to (1) reduce or remove elevation and floodproofing requirements for new and substantially improved agricultural structures, and (2) reduce the cost of flood insurance for agricultural structures. The Task Force recommends that FEMA:

1. Recognize levee relief cuts that are properly planned in an Emergency Operations Plan and adopted by a community—and the resulting lowered Base Flood Elevations (BFEs)--for construction of new and substantially improved agricultural structures and for determining flood insurance premiums.
2. Offer an option to remap Special Flood Hazard Areas as Zone D in leveed areas that meet the following requirements:

   • The community adopts a special floodplain ordinance that requires elevation (or floodproofing) to or above the BFE for new and substantially improved non-agricultural structures in the new Zone D.
   • The community implements a self-reporting program that indicates compliance with the special floodplain management ordinance in the new Zone D.

- The community adopts a levee risk management plan for the new Zone D.
- The community mitigates the loss of the mandatory insurance purchase requirement for the structures in the new Zone D. The Task Force has identified two potential mechanisms for achieving this: (1) an ordinance adopted by the community requiring flood insurance purchase, and (2) flood risk financing by the community (e.g., purchase of a multi-year group flood insurance policy from a private carrier).

3. Revise Operating Guidance 12–13 to map areas behind a certified reach of levee as Zone X (Shaded) instead of Zone D if the certified reach of levee is part of a larger levee system and provides protection from the Base Flood.
4. Allow human intervention for providing entry of floodwaters into wet floodproofed agricultural structures when large doors on at least two sides of the building could be locked open, consistent with an approved Flood Emergency Operation Plan for the structure. Replace the current factor of safety of 5 with 1.5 or an appropriate, technically justified factor of safety for venting of agricultural structures when human intervention is not authorized.
5. Amend insurance rates to reflect the flood protection provided by a non-accredited levee as documented by a civil engineer, following a methodology developed by the Task Force
6. Amend insurance rates to include two separate rating tables for Zone D. One rating table would be for areas identified as 'Zone D Undetermined/Unknown'—the historic Zone D. Create another (new) rating table for areas identified as 'Zone D Protected by Levee'—for areas mapped as Zone D under FEMA's Operating Guidance 12–13.
7. Develop insurance rates for agricultural structures separately from retail business and industrial structures and update the Flood Insurance Manual with the new rates.
8. Recognize wet floodproofing of agricultural structures in insurance rates and address these measures similar to dry floodproofing, updating the Flood Insurance Manual with the new rates.
9. Recognize a subcommunity within a community under the Community Rating System (CRS) program and offer CRS credits for the following activities:

• High ground evacuation locations
• Federal levees with System Wide Improvement Frameworks
• Risk-based levee system improvements
• Levee risk management plans

Federal levees should be eligible for CRS points for levee maintenance, unless the levee is operated and maintained by the federal government. Recommendations 2, 3, 5 and 6 hold the most potential for significantly improving agricultural sustainability in deep leveed floodplains".

3.5 ECOLOGICAL AND SOCIAL BENEFITS OF FLOOD BYPASSES IN THE SACRAMENTO VALLEY

John Cain

3.5.1 Expanded Flood Bypasses and Multi-benefit Flood Management

The CVFPP of 2012 laid the foundation for a number of important flood risk management policies and strategies, but when it came to specific modifications of the actual physical flood management facilities, the plan was fairly conceptual. The 2012 CVFPP proposed expanding three existing flood bypasses: Yolo, Sacramento, and Sutter, and significantly expanding two existing channels into important new flood bypasses: Cherokee Canal in Butte County and Paradise Cut in San Joaquin County (Fig. 3.1). The five bypass expansion projects were the only major physical changes to the system described and mapped in the plan, and even then, the maps of the new bypasses were intentionally fuzzy and of low resolution. The 2012 plan and the debate that ensued around its passage by the new flood protection board did, however, focus flood management planning activities throughout the Central Valley around two important concepts—expanding the conveyance capacity of the system by expanding the bypasses and designing all projects, to the extent feasible, to achieve multiple benefits including ecosystem restoration, open space, recreation, and continued agricultural production. Multi-benefit flood management projects, like expanded bypasses, generally require expanding the designated floodway so that there is enough capacity for flood conveyance, riparian vegetation, and other elements of multi-benefit projects.

The broad outlines of the 2012 CVFPP served as the basis for more specific analyses and proposals for bypass expansion and multi-benefit projects that are included in the 2017 CVFPP Update (CDWR 2017a). New climate change hydrology developed by the California Department of Water Resources (CDWR) and the US Army Corps of Engineers (USACE) for the 2017 plan (CDWR 2017b) bolstered the case for why bypass expansion and new multi-benefit projects that expand the floodway are needed, particularly in the San Joaquin River Basin, where the analysis projects the 100-year flood event to increase by 65 percent by late 21st century in a system that is not currently large enough to reliably convey the historical 50-year event. The same analysis projects the 100-year flood to increase by over 20 percent in the Sacramento Valley by late 21st century. The hydrologic impacts of a warming climate are larger in the San Joaquin River Basin, because it has a far higher median elevation and thus has a much larger area where precipitation will change from snow to rain— effectively increasing the size of the catchment during warm winter storms.

In the Sacramento Basin, the 2017 CVFPP Update focused on incremental expansion of the Yolo and Sacramento Bypasses on the lower Sacramento River near Sacramento, while in the San Joaquin River Basin, the 2017 CVFPP identified several multi-benefit flood management opportunities throughout the catchment with a vastly expanded flood bypass at Paradise Cut. Due to the controversy associated with expanding the Cherokee Canal and Sutter Bypass and the reality that the state does not have the funds or capacity to expand all the Sacramento Valley bypasses at once, DWR staff focused the 2017 Update on the expansion of the Yolo Bypass and deferred any detailed discussions on Cherokee Canal and Sutter Bypass until the 2022 Update.

The greater emphasis on *multi-benefit projects* in the San Joaquin Basin stems from the underlying geomorphic and hydrologic factors that characterize the basin. First and foremost, unlike the Sacramento Valley, where the topography slopes away from the river into vast flood basins, setting back or strategically removing levees along the San Joaquin is not necessarily cost prohibitive and would not flood vast areas in the case of levee removal. Second, the flood system on the San Joaquin is currently so undersized that in many cases expansion of the floodways is the only viable option for preventing unplanned levee failures. Floodway expansion for the sole purpose of flood protection, however, is too expensive to justify along most portions of the San Joaquin, because most floodplains in the San Joaquin Valley are unpopulated. For ecological and recreational purposes, however, there is broad support for financing multi-benefit flood management projects that restore ecosystem function and improve recreational opportunities. Lastly, because of massive reservoirs on the San Joaquin and its tributaries, levees are not necessary to continue seasonal agriculture in 80–90 percent of all years. The reservoirs fully control all runoffs except during the wettest 20 percent of years, greatly diminishing the utility of levees in many places along the San Joaquin River.

The rapidly urbanizing corridor along the lower San Joaquin River in communities of Manteca, Lathrop, and Stockton, however, is one place where levees are necessary to prevent catastrophic flooding of human settlements. Over 30,000 people in the Weston Ranch suburb of Stockton would be deeply flooded if levees protecting reclamation district 17 were to fail in a major flood. A new bypass along the lower San Joaquin River at Paradise Cut is specifically planned to divert flood waters away from this rapidly urbanizing corridor and into the undeveloped agricultural lands of the south delta.

Paradise Weir, a small rock structure at the head of Paradise Cut, currently diverts floodwaters out of the mainstem and away from these urbanizing areas, but its current capacity is extremely limited. The expansion plan calls for adding a new 300-m (1000 feet) wide weir 2 miles upstream of the existing weir and adding over 809 ha (2000 acres) of farmland to increase the capacity of the area of the bypass threefold. The proposed bypass will lower flood stage by 7.6 m (2.5 feet) at the I-5 Bridge, where there is significant new and planned urban development. Furthermore, it will lower flood stage along 48 km (30 miles) of river extending from the Port of Stockton to the confluence with the Stanislaus River.

Paradise Cut is intended to protect urban communities along the San Joaquin River in the same way as the Yolo Bypass protects urban developments on the lower Sacramento River (Fig. 3.3). The Yolo Bypass, which for the purposes of this description includes the Sacramento Bypass, was originally completed in the 1920s to divert water away from Sacramento and into an undeveloped flood basin in the north delta. Flood waters enter the Yolo Bypass at Fremont Weir near the confluence of the Feather and Sacramento Rivers. The weir is a simple concrete structure 2.1 m (7 feet) tall and 2.6 km (1.6 mi) long, which allows water to flow out of the Sacramento River and into the undeveloped Yolo Basin, where it then drains slowly to the tidal Sacramento-San Joaquin Delta. Near the confluence of the Sacramento and American Rivers, the manually operable Sacramento Weir allows floodwaters from the American River to flow into the Yolo Bypass during extreme events. During large floods, the Bypass is designed to carry 14,158.4 m³/s (500,000 ft³/s), nearly five times the design capacity of the lower Sacramento River past downtown Sacramento.

The 2017 CVFPP Update includes two major projects to expand the Yolo Bypass—the lower and upper Elkhorn Basin projects. Depending on the exact configuration, the combined projects will cumulatively reduce flood stage for the 200-year event by 0.76–1 m (2.5–3 feet) along densely urbanized reaches along the Sacramento River. Furthermore, the combined projects would reduce flood stage along more than 193 km (120 miles) of the Sacramento River both upstream and downstream of the bypass.

In addition to modifications to the Yolo Bypass for flood stage reduction, the state and federal governments are also in the midst of planning significant modifications to Fremont Weir and the bypass to provide adult fish passage for endangered salmon and sturgeon and to create frequently inundated floodplain habitat for juvenile salmon and other native fish and wildlife. These improvements are purely for the purposes of improving habitat and will do nothing to reduce flood risk. Very importantly, however, expansion of the bypass will not only reduce flood risk but also significantly facilitate opportunities for expanding habitat function in the

Fig. 3.3 Yolo Bypass floodplain and view of downtown Sacramento, 2007 (Courtesy of California Department of Water Resources)

bypass. It is very difficult to add habitat features such as trees to a floodway without increasing flood stage or velocities, but it is significantly easier to add habitat features as part of a bypass expansion effort designed to both lower flood stage and improve habitat. The State of California has made a reasonable effort to integrate these two separate efforts, but varying time-lines and objectives combined with the inherent challenge of coordinating multiple state and federal agencies has made the task of integration exceedingly difficult.

The Lower Elkhorn project expands a portion of the Yolo Bypass as well as the Sacramento Bypass, a tributary to the Yolo Bypass that directs American River water flowing over Sacramento Weir into the larger Yolo Bypass. The project will widen the Sacramento Bypass by 457 m (1500 feet) to enable it to convey more water into the Yolo Bypass significantly reducing flood stage on both the American and Sacramento Rivers, both of which pose a significant flood risk to the City of Sacramento. The lower Elkhorn project will not widen or replace the Sacramento weir, which controls flow into the Sacramento and Yolo Bypasses because of the engineering complexities of rebuilding or rerouting a rail bridge that runs across the existing weir. But widening the bypass in the absence of widening the weir is still expected to lower flood stage by as much as 1 foot. Future expansion of the Sacramento Weir by the Corps of Engineers as part of the Sacramento River Common Features project will further lower flood stage and will provide an opportunity to improve fish passage over the weir. The Lower Elkhorn project will also set back 9.3 km (5.8 mi) of levee on the east side of the Yolo Bypass by 457 m (1500 feet) immediately upstream of the Sacramento Bypass to expand a constrained area of the existing bypass. As of 2017 the state had reserved $200 million in funding for the implementation of the Lower Elkhorn Basin project and was busily working on completing design and environmental compliance documents.

The Upper Elkhorn project significantly expands both the upper Yolo Bypass and Fremont Weir. It will extend Fremont Weir by 1 mile and set back an 8 km (5 miles) segment of the east levee by 2.4–3 km (1.5–2 miles) greatly expanding the ability of the Yolo Bypass to reduce and redirect peak flows off of the Sacramento and Feather Rivers. Depending on how it is ultimately designed, the Upper Elkhorn Bypass could significantly increase the area of seasonally inundated floodplain for juvenile salmon, migratory waterfowl, and other native species. The Upper Elkhorn Basin project is significantly more expensive and will take longer to complete because it entails conveyance improvements to the lower bypass that must occur before the upper bypass can be safely expanded. The state has not yet identified any funding sources to complete the Upper Elkhorn project.

The proposed bypasses will create flexibility for a broad range of ecosystem restoration activities by freeing up capacity within the bypass and the bypassed river reaches as well as the river channels upstream of the bypasses. The expansion of all five bypasses will eventually include several habitat restoration and protection elements including fish passage facilities over Fremont and Sacramento Weirs for anadromous fish; riparian vegetation along the low flow channels in the bypasses; and seasonal floodplain wetlands for migratory birds and native fish including Chinook salmon. Furthermore, the bypasses improve connectivity between rivers and their floodplains and restore more natural hydraulic and geomorphic processes upstream and within the bypassed reaches.

3.5.2   Multi-benefit Flood Management Projects

Levee setback projects that would expand the floodway and provide other benefits such as habitat restoration were not initially included in early drafts of the 2012 plan, but after pleas from environmental interests, "strategic levee setbacks" were included in a list of generic measures identified as consistent with the plan. DWR engineers did not embrace widespread levee setbacks in the 2012 plan due to the expense and limited flood risk reduction benefits. Because flood stage is generally controlled by the cross-sectional area of the downstream channel, local levee setbacks only had localized and limited flood stage reduction benefits much the way as expanding the middle of a garden hose would not increase the amount of water that could be pushed through the hose. Instead, the plan focused on expanding flood bypasses in the lower end of both Sacramento and San Joaquin levee systems, where most of the urban areas at risk of flooding are concentrated. Unlike local levee setbacks elsewhere, expansion of the floodways at their downstream ends can result in flood stage reduction for many kilometers upstream.

The Board resolution for the 2012 plan did, however, call for multi-benefit projects "wherever feasible," and subsequent developments have resulted in a proliferation of multi-benefit flood management project proposals. The exact origins of the term "multi-benefit flood management" in the context of the Central Valley is not clear, but it has galvanized political support around a different type of flood management. During the debate over the 2012 plan, all parties agreed that financing flood system improvements in the Central Valley was going to be an uphill battle. Environmental stakeholders successfully argued that the wealthy coastal areas of California were far more likely to fund flood management projects in the Central Valley if those projects delivered benefits valued by coastal populations such as improved fish and wildlife habitat. In the aftermath of the 2012 flood plan, several environmental organizations developed a website defining multi-benefit flood management with the following definition and showcasing a number of specific projects.

Multi-benefit Flood Protection Project "Multi-benefit projects are designed to reduce flood risk and enhance fish and wildlife habitat by allowing rivers and floodplains to function more naturally. These projects create additional public benefits such as protecting farms and ranches, improving water quality, increasing groundwater recharge, and providing public recreation opportunities, or any combination thereof" (American Rivers 2017).

Environmental groups subsequently lobbied to get the term "multi-benefit flood management" incorporated into a major water bond placed on the ballot in 2015, and voters approved the bond, which included over $400 million for multi-benefit flood management projects. In the spring of 2016 the legislature directed DWR to prioritize remaining expenditures from a previous flood control bond from 2006 for multi-benefit flood management projects. A few non-governmental organizations interested in promoting multi-benefit projects successfully worked with local levee districts to obtain state and federal funding for several multi-benefit flood management projects or persuade DWR to include them in the 2017 plan. Examples include breaching the levee at the San Joaquin River National Wildlife Refuge to reconnect 1200 ha (3000 acres) of floodplain, restoring 809 ha (2000 acres) of riparian and floodplain habitat at the Dos Rios Ranch across the river from the refuge, reconnecting the Oroville Wildlife Area to the Feather River to create rearing habitat for juvenile salmon, setting back 11 km (7 miles) of levee on the upper Sacramento River to reconnect 566 ha (1400 acres) of floodplain and improve flood protection for Hamilton City, expanding the floodway on the upper San Joaquin River to provide 100-year protection for the City of Firebaugh along with wildlife habitat and recreational opportunities for an economically disadvantaged community.

The Southport Project in the City of West Sacramento is one of the most exciting and ambitious multi-benefit floodplain projects moving forward in the Central Valley. West Sacramento has a population of 52,000 people and is situated on the west side of an urbanized reach of the Sacramento River that is severely constrained by levees. The project is necessary to provide 100-year protection for the established and growing urban center. Due to very deep sandy soils, it was not feasible to fix the existing levee in place with a seepage curtain. Engineers determined that setting back the levee and constructing a massive seepage berm on the landward side of the levee was the best way to protect the community. The project will set back 9 km (5.6 miles) of levee creating a new floodplain area 121–304 m (400–1000 feet) wide creating 80 ha (200 acres) of frequently inundated floodplain habitat in the heart of the Sacramento metropolitan area. The project will ultimately provide high-quality habitat for several native species including juvenile rearing habitat for endangered Chinook salmon.

3.5.3   Conclusion and Challenges

Reforms over the last decade have laid the foundation for a new era of multi-benefit flood management in California's Central Valley as described in the 2012 CVFPP and the subsequent 2017 Update. Initial public investments will focus on expanding the flood bypass system first proposed in the late nineteenth century and constructed in the early twentieth century. Expansion of the bypass system is essential for protecting urban populations against the larger floods that a warming climate is expected to generate in California's Central Valley. Expanding the bypasses will also increase opportunities for incorporating a number of other benefits into the flood management system including habitat restoration and recreational amenities. Independent of the bypasses, public enthusiasm for habitat restoration and outdoor recreation has resulted in the development of several new multi-benefit flood management projects throughout the Central Valley which will breach levees and restore floodplain habitats.

Implementation of the bypass expansion projects and other multi-benefit restoration projects described in this chapter will require overcoming major funding and permitting constraints. Any modification to the existing flood system will require navigating the Corps of Engineers' cumbersome section 408 permitting process pursuant to section 14 of the Rivers and

Harbors Act of 1899 (33 USC 408). In 2014, the Corps of Engineers provided new guidance on the 408 permit process (USACE 2016). Unless state and local project sponsors are able to develop efficient strategies for completing successful 408 applications, several good multi-benefit flood management projects could stall indefinitely. Obtaining regulatory permits under the federal Clean Water Act and the state and federal Endangered Species Acts is another major hurdle to efficient implementation because nearly all projects involve disrupting waters of the USA that harbor endangered species.

Lack of funding to implement the bypasses, other multi-benefit projects, and other elements of the CVFPP is another major hurdle. The state has already dedicated some general obligation bond funding approved by the voters to fully finance some multi-benefit projects and partially finance implementation of the bypasses. Both DWR and the CVFPP hope to identify new funding sources for both implementation and maintenance that are not dependent on periodic bond measures approved by the voters. The 2017 CVFPP identifies some potential long-term, stable funding sources including a statewide river basin assessment, a valley-wide drainage district assessment, or a new state flood insurance program. But all of these funding sources would require approval of the legislature or voters.

Streamlined procedures to efficiently account for how varied funding sources are legally used to finance various aspects of multi-benefit projects are another major financing challenge. Many multi-benefit flood management projects cannot be financed by a single special district with a limited, special-purpose charter such as flood control or water quality, because these agencies are legally prohibited from spending funds on purposes outside of their charter. These projects could be financed by a mix of different funding sources reserved for special purposes like parks, recreation, water quality, flood management, and habitat enhancement, but deciding how much of any one project can or should be paid for with any pool of special-purpose money is generally dictated by the unique requirements of different funders making true integration difficult. For example, some park and trail districts don't want to pay for any facilities (i.e., trails or active park lands) below the 100-year water surface elevation, while some flood management agencies believe that they are not able to spend any money on habitat that is not explicitly required by environmental mitigation requirements. The challenge of integrating these varied and sometimes conflicting preferences and requirements into an efficient financing plan has made it very difficult to finance implementation with a package of disparate funding sources. The transition from single-purpose flood control projects to multi-benefit flood management under the new paradigm of integrated water management will ultimately require new accounting mechanisms for financing integrated water management.

REFERENCES

American Rivers. 2017. *A Multi-Benefit Approach to Flood Protection in California's Central Valley.* http://www.multibenefitproject.org/what-is-multiple-benefit/. Accessed 3, 5 Sept 2017.

CDWR (California Department of Water Resources). 2005. *Flood Warnings: Responding to California's Flood Crisis,* January. http://www.water.ca.gov/pubs/flood/flood_warnings___responding_to_california%27s_flood_crisis/011005floodwarnings.pdf. Accessed 9 Sept 2017.

———. 2010a. *State Plan of Flood Control Descriptive Document,* November.

———. 2010b. *A Handbook for Local Communities,* October.

———. 2011. *Flood Control System Status Report,* December.

———. 2012a. *2012 Central Valley Flood Protection Plan,* June. http://www.water.ca.gov/cvfmp/2012-cvfpp-docs.cfm. Accessed 9 Sept 2017.

———. 2012b. *Urban Levee Design Criteria,* May.

———. 2017a. *Central Valley Flood Protection Plan Update: Climate Change Analysis Technical Memorandum,* March. Sacramento. http://www.water.ca.gov/cvfmp/docs/CC_DraftClimateChangeSummary_March2017.pdf.

———. 2017b. *Central Valley Flood Protection Plan Update: Climate Change Analysis Technical Memorandum,* August. Sacramento.

FEMA (US Federal Emergency Management Agency). 2017. *Special Flood Hazard Area.* https://www.fema.gov/special-flood-hazard-area. Accessed 24 Aug 2017.

HDR, Inc. 2016. Technical Memorandum, Agricultural Floodplain Ordinance Task Force. *Recommended Administrative Refinements of the National Flood Insurance Program to Sustain Agriculture as a Wise Use of the Floodplain in Leveed Special Flood Hazard Areas,* December 28.

Kelley, R. 1998. *Battling the Inland Sea.* Berkeley: University of California Press.

Lund, J.R. 2012. Flood Management in California. *Water* 4: 157–169. https://doi.org/10.3390/w4010157.

Senate Bill 5 (Machado). 2006–07. *Session of California State Legislature.*

US Department of the Interior/ US Geological Survey. 2011. California Water Science Center, June.

USACE (US Army Corps of Engineers). 2016. EC 1165-2-216. Procedural Guidance for *Processing Requests to Alter US Army Corps of Engineers Civil Works Projects Pursuant to 33 USC 408.* Washington, DC 20214-1000: Department of the Army.
————. Sacramento District. http://www.spk.usace.army.mil/Missions/Civil-Works/Sacramento-River-GRR/. Accessed 10 Aug 2017.

CHAPTER 4

Managing Floods in Large River Basins in Europe: The Rhine River

Laurent Schmitt, Dale Morris, and G. Mathias Kondolf

Abstract While not among the globe's largest rivers in terms of discharge, the Rhine has assumed outsized importance due its role in history and as the spine of Western Europe, with an extraordinary concentration of population and industry, and consequently profound anthropic modifications. Floods have always plagued the Rhine, but twentieth-century corrections of the river increased the flood risk by taking away the river's natural floodplain storage areas. These are being restored along the Upper Rhine with multiple projects to reconnect bypassed sections of the river and allow frequent flooding of polders, projects that not only reduce flood risk but also restore habitat and water quality. Similarly, in the Rhine Delta, the celebrated "Room for the River" projects have taken back some of the river's floodplain areas as a way to accommodate flood flows and restore lost habitats.

L. Schmitt
University of Strasbourg, Strabourg, France

D. Morris
Embassy of Netherlands, Washington, DC, USA

G. M. Kondolf (✉)
University of California Berkeley, Berkeley, CA, USA

University of Lyon, Lyon, France

© The Author(s) 2018
A. Serra-Llobet et al. (eds.), *Managing Flood Risk*,
https://doi.org/10.1007/978-3-319-71673-2_4

75

Keywords Rhine River • Restoration Upper Rhine • Room for the River • Rhine Delta • Floods Directive

4.1   INTRODUCTION

Laurent Schmitt G. Mathias Kondolf

4.1.1   The Rhine River Basin and Its Significance

With a basin covering an area of about 185,000 km² and flowing through Europe over 1250 km from its sources to the North Sea, the Rhine is the third largest river of Europe and in many ways the "spine" of the original European Union. The Rhine has been a major vector of economic (e.g., navigation, power generation, industrial production, municipal water supply) and cultural development for over two millennia and now has a population of 58 million inhabitants, concentrated mostly in large cities along the river itself. By virtue of the number of countries in its basin, the economic and political importance of the Rhine axis, and its history of anthropic modifications, the Rhine is a particularly compelling case study of river basin management in general, and most notably of recent and ongoing national and international efforts to manage flood risk in creative ways that also produce ecological and social benefits, supported by interdisciplinary research.

In this chapter, Laurent Schmitt and Matt Kondolf provide an overview of the Rhine River—its physical geography and recent history; Schmitt then describes the human alterations to the Upper Rhine, the urgent need to increase flood capacity in this highly modified part of the river, and initial projects that have achieved both flood risk reduction and restoration of ecological function; Dale Morris then reports on the widely hailed "Room for the River" (RfR) efforts in the Netherlands, wherein additional flood capacity is created while restoring ecological function within the Rhine Delta.

4.1.2   Physical Geography of the Rhine

The basin drains part of the Northern Alps, nearby mountain ranges (e.g., Black Forest, Vosges, and Jura), and lowlands from the foothill of the Alps to the North Sea. The river drains a diverse geology, ranging from granitic rocks in the headwaters, and many other lithologies such as limestone, graywackes, gneiss, sandstone, and schists, as well as unconsolidated Pleistocene deposits. The geological history of the Rhine is complex: pre-Pliocene, it was tributary to the eastward-draining River Donau, then it flowed westward toward the River Rhône (part of the Alpine Basin), and since the Pleistocene, it adopted its current northward course (CHR/ KHR 1977).

The longitudinal profile of the Rhine is controlled by two local base levels: the Lake Constance (Bodensee) where the sediments of the Alpine Rhine are deposited, and the upstream extremity of the narrow bedrock valley of the Middle Rhine. The Rhine Basin and Rhine Valley are classi-cally subdivided into six reaches (Fig. 4.1) (Uehlinger et al. 2009): the Alpine Rhine, down to Lake Constance, the Hoch Rhine (or High Rhine) from the Lake Constance to Basel, the Upper Rhine, a 350-km reach from Basel to Bingen through a large tectonic graben, the bedrock-controlled Middle Rhine from Bingen to Koblenz, the Lower Rhine from Koblenz to the beginning of the delta, and finally the Rhine Delta (coalescent with the Meuse Delta) (see Chapter 4.3), where the Rhine divides into three main branches (Ijssel, Neederijn-Lek, and Waal).

Mean temperatures range from 8.5°C to 11°C, and basin-wide annual precipitation averages about 950 mm (up to 2000 mm in the Alps) (Uehlinger et al. 2009). The average discharge at the mouth is about 2300 m^3/s, of which nearly half (1030 m^3/s) is derived from the upstream-most 20% of the basin, above Basel, illustrating the strong influence of the Alps on the Rhine. Upstream of the River Main confluence (Upper Rhine), the Rhine's hydrological regime is nival-glacial, with high flow during June and July and low flow during January and February. Downstream, all tributaries have pluvial regimes (high flow during February and March and low flow during August and September). The increased winter flows are most pronounced in the lower reach of the Rhine. Large lakes upstream of Basel reduce the variability of the Rhine hydrological regime (Uehlinger et al. 2009).

4.1.3   Floods, Water Quality, and River Basin Management

The intensive use of the Rhine resources going back centuries has pro-duced large alterations, especially along the Upper Rhine (see Sect. 4.2). The nineteenth-century "corrections" by the engineer Tulla, followed by further works in the twentieth century, resulted in an overall shortening of

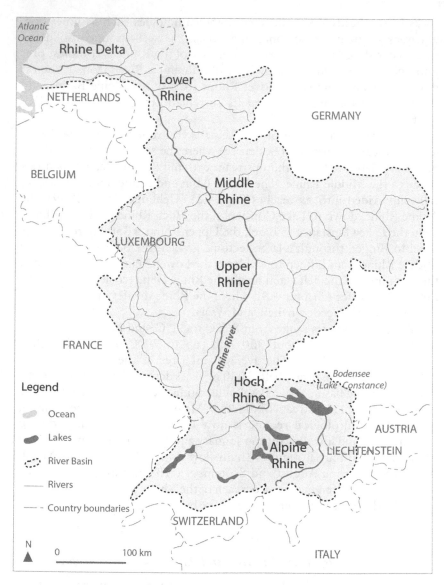

Fig. 4.1 The Rhine River Basin

80 km from Basel to Hesse, and consequent channel incision, drops in alluvial groundwater levels, increased flood stages (and thus greater flood risk) in the Middle Rhine and downstream, and loss of important channel and floodplain habitats (Wieriks and Schulte-Wulwer-Leidig 1997). With both population centers and industrial areas concentrated along the Rhine, water quality has long been a concern, as manifest in the disappearance of salmon from the system during the 1960s and by the establishment of the International Commission for the Protection of the Rhine (ICPR) in 1950. The notorious spill of pesticides and other chemicals in late 1986 from the Sandoz Chemical complex near Basel put water quality of the Rhine in front and center of attention in Europe and, along with extensive algal blooms in the North Sea off the mouth of the Rhine, spurred concrete action by the ICPR and the environment ministers of the riparian countries; adoption of the Rhine Action Programme signaled an international commitment not only to cleaning up water quality but also to restoring the river's ecosystem (Wieriks and Schulte-Wulwer-Leidig 1997).

Catastrophic floods in 1993 and 1995 inspired a broadening of the ICPR role to include flood management and structural reorganization to accommodate this expanded role. The Rhine experience in international water quality and flood management was an important and influential precursor to the Water Framework Directive (adopted by the EU Parliament in 2000) and the Floods Directive (adopted in 2007) (Mostert 2009).

In this chapter, Laurent Schmitt describes projects to reconnect floodplains cut off from the Upper Rhine in the twentieth century to restore flood capacity and ecosystem function. Dale Morris reports on similar projects in the Rhine Delta as part of the celebrated RfR program.

4.2   Restoring Flood Capacity and Ecological Function Along the Upper Rhine

Laurent Schmitt

4.2.1   The Upper Rhine Before Significant Human Alterations

As the Upper Rhine drops from 246 m to 89 m in elevation from Basel to Bingen, there is a gradual decrease in gradient, and the natural channel pattern changed from braided, to anastomosing, to meandering. These changing channel patterns exhibit high geomorphological diversity, and hence high biodiversity (Carbiener 1970; Carbiener and Schnitzler 1990). In the reach Neuf Brisach to Strasbourg, the Upper Rhine was aggrading during the Holocene (less than 1 m; Schmitt et al. 2016), when it was incising elsewhere, resulting now in the division of the flow into up to 18 channels of different types, with unusually high ecological diversity (Carbiener and Schnittzler 1990). Consequently, the Upper Rhine was also characterized by a great abundance of fishes, especially salmon (Uehlinger et al. 2009), which were present throughout the Rhine downstream of the Schaffhouse waterfall, and which supported an intensive fishery. Fishing, navigation, and other activities linked to the Rhine thrived despite the permanent risk of floods, which historically induced extensive damages, especially in the wide floodplain downstream Neuf Brisach (Wetter et al. 2011).

Between the twelfth century and the end of nineteenth century, the Upper Rhine experienced at least six major floods exceeding 6000 m³/s and about 40 floods ranging between 5000 and 6000 m³/s. The biggest flood, called "Le Déluge du Rhin" in France, occurred in August 1480 and inundated the entire alluvial plain (Wetter et al. 2011).

4.2.2   *Three Phases of Engineering Works Since the Nineteenth Century*

Since the beginning of the nineteenth century, three successive engineering works modified the functioning of Upper Rhine fluvial hydrosystem drastically and irreversibly:

The *correction* (1817–1876), planned by the German engineer Johann Gottfried Tulla, stabilized the main channel between two artificial banks and limited the floodplain between two high flow dikes. The aims were flood control, border fixation, agriculture and forest development, and improvement of human health. The main channel was narrowed from about 1000 m to 250 m, many meander bends were cut off (increasing slope), and banks were stabilized. The corrected channel incised about 1 m, except from Neuf Brisach to Strasbourg, where the longitudinal profile was almost stable. The bed incised up to 7 m above Neuf Brisach, forming an armor layer. At Istein, the channel incised through bed gravels and exposed underlying bedrock, which formed rapids and impaired navigation, especially for the Basel harbor. In this incised upstream reach, all lateral channels disappeared (CHR/KHR 1977). Downstream, braiding channels disappeared or were transformed into (new) anastomosing channels, when the remaining flooding areas showed general deposition of fine sediments.

The *regularization* (1930–1962) involved construction of in-channel groin fields to improve navigation, which induced an addition of approximately 1 m incision.

The *canalization* (1932–1977) involved construction of a series of five artificial canals constructed on the western (French) side, as stipulated in the "Traité de Versailles" following the First World War. The upstream canal, from Kembs to Neuf Brisach, the "Grand Canal d'Alsace," supports four power plants. From Neuf Brisach to Strasbourg, each bypass supports one power plant. Downstream of Strasbourg, the Rhine bed itself was concretized and it supports two power plants. An 11th power plant planned for Neuburgweier was not built. The main aims of the canalization were both power generation and improvement of navigation. It also favored industrialization along the Rhine. The five bypassed reaches, referred to as the "Old Rhine," were dewatered except for a small minimum flow release, and during floods when the canal capacity of about 1500 m³/s is exceeded and excess flows are rejected into the Old Rhine (Staentzel et al., in revision). The canalization significantly increased flood risk downstream by eliminating lateral connection with 130 km² of floodplain that formerly stored floodwaters (CHR/KHR 1977; Dister et al. 1990). It also disconnected lateral channels and produced severe ecological losses. Downstream of Iffezheim (the downstream-most power plant), an average of 170,000 m³ of sediment is injected annually into the Rhine since 1977 to control bed incision (CHR/KHR 1977; Kuhl 1992).

4.2.3   Restoring Flood Protection and Ecological Function: A Fourth Increasing Phase of Ecohydrological Engineering Works

As the effects of the canalization became apparent (CHR/KHR 1977, Commission d'Etude des Crues du Rhin 1978), an international convention committed to restoring pre-canalization levels of flood protection (Convention 1982; Commission Permanente 2016), by retaining 272.6 million of m³ in the 200-year flood, mostly in Germany (216 m³). Eighteen flood retention areas ("polders") have been constructed or are planned, of which two are in France. Other measures consist of operational changes for power plants, closing two agricultural dams and lowering (6 m) the areas of floodplain in the upper reach along the German side of the Old Rhine over a length of about 40 km and a width of less than 100 m.

In Baden-Württemberg, the Integrated Rhine Program includes this floodplain lowering as well as the construction of 10 polders (Brendel and Pfarr 2017). These areas are also subject to frequent "ecological flooding" and restoration of lateral connectivity of side channels, as also implemented in the Erstein polder in France (Trémolières et al. 2008, see Fig. 4.2). Considering that hydrological and ecological objectives are not mutually exclusive, but rather offer mutual benefits (Dister et al. 1990), the approach of restoring flood capacity by reconnecting floodplain areas along the Upper Rhine (which has successfully reduced flood risk) has also provided tremendous opportunities for ecological restoration and improvement of water quality.

Despite the impacts of engineering works, the Upper Rhine hydrosystem still shows an important biodiversity which motivated the protection of large areas in designations such as natural reserves, a trans-boundary Ramsar site, and initiation of more than 120 restoration actions over three decades (Schmitt et al. 2012; Schmitt and Beisel 2015). These efforts intensified after the Sandoz accident in 1986, after which the water quality was also significantly enhanced. Restoration actions consist of lateral channel reconnection, construction of fish passes, channel creation, floodplain lowering, increased instream flow releases and restoration of high flows as

Fig. 4.2 Aerial picture of the Erstein polder during the 2004 retention flood, view in the southern direction (Courtesy of Voies Navigables de France)

ecological floods (Staentzel et al. under review), gravel augmentation (Arnaud et al. 2017), and controlled bank erosion. Despite these extensive efforts, restoration remains highly constrained by engineering structures and hydrological limitations on the canalized section (mostly on the French side); many channel reconnections proved to be too small to produce significant results.

In a few cases, the restored sites have been monitored in order to evaluate if the restoration objectives are obtained and to assess the trends of recovery (Schmitt et al. 2012; Schmitt and Beisel 2015). In many cases, monitoring has been conducted only on certain species, without accounting for abiotic driving factors, such as hydrology and morphodynamics. A recent international conference in Strasbourg concluded that shared abiotic and biotic metrics were needed for monitoring and that restoration projects should be bigger and target more processes, especially the hydrological and morpho-sedimentary dynamics that control ecological processes. Another insight from this conference was that the large number of actors involved made it difficult to have an overview of all the restoration efforts and to share lessons learned.

An international observatory for the restoration of the Upper Rhine, whose elaboration is in progress (Schmitt and Beisel 2015), should promote an international view of the restoration of the Rhine River and its floodplain, and support comparison of monitored results and creation of a shared data base, to better accumulate and exchange experiences to improve future restorations and to rehabilitate more effective and sustainable fluvial processes (Schmitt and Beisel 2015). For example, the restoration measurements in the Erstein polder, which included re-flooding and channel reconnection, included a scientific monitoring with an interdisciplinary approach over a relatively long period of six years (Trémolières et al. 2008). On this basis, improvements in restoration approaches are currently considered. Preliminary outputs of this observatory also show that while the first restoration projects were mostly small scale and not followed by environmental monitoring, more recent projects are generally more ambitious, in terms of spatial extent and natural processes which are restored, and better monitored.

The complementary restoration of flood retention and ecological function on the Upper Rhine currently in progress has shown positive results, but much remains to be done. The efforts are costly, which requires that in the future they be designed in the most sustainable manner possible.

4.3   Room for the River in the Rhine Delta

Dale Morris

4.3.1   Room for the River: Context

The RfR program is a multi-year, multi-project effort to increase the design discharge levels in the Rhine River and in the Maas and Scheldt river distributaries that form the Rhine River (Dutch) Delta in the Netherlands. *Rijkswaterstaat,* the national government's water management agency, was primarily responsible for RfR, with support for the regional water authorities; municipal and provincial governments were also involved, primarily to ensure project compatibility with spatial planning goals and policies. Project costs were shared between the parties.

RfR projects were engineered/constructed between 2009 and 2016 and were completed on time and within budget (2.3 billion euros) (Fig. 4.3). RfR had two primary and co-equal goals: to reduce riverine flood risk and improve the spatial and environmental quality of project

Fig. 4.3 River widening and waterfront redevelopment at Nijmegen/Lent, the Netherlands. View from 2015 showing project nearing completion, with broader river channel, and diversion channel, new waterfront, road, bridge and rail infrastructure (Courtesy of Johan Roerink and H+N+S Landscape Architects)

area. Some 82% of residents in the project areas agree that flood protection levels are higher than before the projects were built, and 84% believe that the projects produced more attractive river areas. The dual requirements to *reduce flood risk and improve spatial quality* requirement is perhaps the key for RfR's high acceptance.

Almost 60% of the Netherlands is at or below sea level and 60% of the Dutch populace lives in, and 70% of Dutch GDP is produced in, flood-prone areas. Robust flood protection is the only way Dutch citizens can safely occupy the dangerous, yet productive delta landscape. Because the Netherlands is subject to both coastal and riverine floods, Dutch history is best illustrated by repeated, devastating floods leading to yet another round of dike heightening and land reclamation measures.

The last major coastal flood occurred in January 1953, which motivated a 40-year effort—known as the Delta Works—to shorten the Dutch coastline and strengthen coastal defenses. Just as Delta Work projects were nearing completion, two major near-floods in the eastern Netherlands in 1993 and 1995, focused attention on the Dutch river landscape. While upstream encroachments had substantially constricted floodplains, river dike heightening and strengthening over the previous century had severed the rivers' access to their natural floodplains and had encouraged more human settlements "behind the dikes." Not only were more people at risk, the rivers' natural ability to absorb floods was substantially reduced. Climate scientists were projecting more extreme weather—wetter winters, drier summers—over northern Europe in the coming decades. RfR is perhaps the Netherlands' first national "climate adaptation" policy.

4.3.2   Room for the River: Policy and Implementation

Conceptually, RfR is best summarized as a conceptual approach: Retain, Store, Drain. Retain water where it falls, Store water temporarily when needed, Drain water when necessary. Beginning in the 1980s, given the growing resistance to river dike heightening, Dutch engineers developed a set of ideas—river widening, bypasses, diversions, groin elimination, dike setbacks, and so on—to help land-use planners and flood managers redesign the river landscape and increase the "design flood" discharge capacity by about 30% (from 12,500 cm/s to 18,000–20,000 cm/s). Enhanced discharge capacity protects upstream areas during periods of maximum discharge and downstream areas in periods of high tides, also taking into account anticipated higher mean water levels in the Dutch Delta.

The integrated RfR policy and programs were developed between 1996 and 2006, and constructed between 2009 and 2016. Between 2006 and 2009, a crucial stakeholder engagement process encouraged citizens to help select specific "Room" projects. A "software toolbox" of 700 possible measures—large, small, traditional, non-traditional—was available, which enabled all parties to explore and design RfR projects and then discuss them with planning officials at local forums.

The toolbox substantially enhanced the public's overall understanding of RfR's goals and educated citizens about how upstream and downstream flood risk reduction and river discharge measures interact. During this engagement process, citizens were also able to share their goals for "added community benefits" that could be achieved through the RfR project. Of the 700 possible measures identified, 39 specific projects were selected in 2009, and 34 were eventually constructed.

The stakeholder meetings were led by local officials and supported by *Rijkswaterstaat* and independent experts. Because RfR also mandated spatial quality improvements to the project areas, environmentalists, cyclists, nature lovers, city planners, and developers took part in the discussions. Given the robust community engagement, RfR implementation was smooth and mostly non-controversial. An overview of RfR is here: https://www.ruimtevoorderivier.nl/english/.

4.3.3   Room for the River: Inspiration for the USA?

After the 2008 and 2011 Mississippi River floods, the 2011 Vermont and 2015 Charleston floods, many US communities see RfR as highly relevant to their future. RfR projects are innovative, worthy of study, and arguably, of replication in the USA and elsewhere. Some RfR projects, however, are not the "least costly option," because they combine multiple strategies to achieve multiple goals: flood risk reduction, floodplain improvements, environmental restoration, and economic/waterfront development. Combining such diverse functions and goals as flood risk reduction and improved spatial quality in the same project clearly facilitated design, acceptance, and implementation. Mirroring that process via an integrated project across US federal, state, and local actors is difficult to imagine.

Other important differences between the Netherlands and the USA may also limit RfR applicability in the USA. First, a general understanding of flood risk is prevalent in the Dutch national psyche, and many citizens and most policymakers know that a major flood catastrophe in the Netherlands could devastate the entire country and its economy for years. This awareness, and the threat of widespread economic damage, enables the Dutch to overcome policy stasis more easily. Flood risk reduction efforts are not a matter of "if" in the Netherlands but of "how." Conversely, floods along the Mississippi River, or in California's Central Valley, impact the USA national psyche and national economy to a much lower degree. Floods in the USA are mainly regional or local issues, and even when the economic impacts are national, they are not overwhelming. Thus, building support for a US RfR approach might be much harder, given that impacts and salience are mostly localized.

Discharge levels, water flows, and water-level fluctuations are much more dynamic along the US rivers than along many Dutch rivers. US catchments often have more diverse settlement patterns, ecosystems, biology, and land uses, and are larger. RfR-type projects may therefore need to be more robust along these US rivers, which increases the difficulty of planning, designing, funding, and implementing such interventions. Moreover, the Dutch political system is primarily hierarchical, with the national government in a dominant position. While provincial and municipal governments, as well as the powerful Dutch Water Boards, have crucial roles, the top-down nature of Dutch flood protection efforts facilitates implementation. The US federal system, in which states and local governments have considerable autonomy, and where various authorities are diffused between federal agencies, makes project design and implementation more complex (as illustrated along the Mississippi, see Sect. 2.3).

An issue which gains much attention in the USA-Netherlands RfR discussion concerns private property. Land in Dutch floodplains—as in the USA—is mostly intensively farmed and privately held. Navigation and logistics along Dutch rivers support the national and regional economies. Rivers and floodplains are home to endangered, protected, and migratory species, and are used for recreation. RfR projects have impinged upon these uses, and thus negotiated easements, and expropriation of private property occurred during implementation. Because private property rights are well established and strongly enforced in the Netherlands, the Dutch government had to make accommodations for impacted property owners. Fair-market compensation, land purchases/exchanges, independent arbitration, resettlement—these were approached openly and constructively as part of RfR.

REFERENCES

Arnaud, F., H. Piégay, D. Béal, P. Collery, L. Vaudor, and A.J. Rollet. 2017. Monitoring Gravel Augmentation in a Large Regulated River and Implications for Process-Based Restoration: Monitoring Gravel Augmentation in a Large Regulated River. *Earth Surface Processes and Landforms.* https://doi.org/10.1002/esp.4161.

Brendel, M., and U. Pfarr. 2017. Integriertes Rheinprogramm: viel Ökologie im Hochwasserschutz. *Bundesverband Beruflicher Naturschutz e.V. (Hrsg.). Jahrbuch für Naturschutz und Landschaftspflege* 61: 128–133.

Carbiener, R. 1970. Un exemple de type forestier exceptionnel pour l'Europe occidentale: la forêt du lit majeur du Rhin au niveau du fossé rhénan (Fraxino-Ulmetum).Intérêt écologique et biogéographique. Comparaison à d'autres forêts thermophiles. *Vegetatio*: 96–148.

Carbiener, R., and A. Schnitzler. 1990. Evolution of Major Pattern Models and Processes of Alluvial Forest of the Rhine in the Rift Valley (France/Germany). *Vegetatio* 88: 115–129.

CHR/KHR. 1977. Le bassin du Rhin. Monographie. *Commission Internationale de l'Hydrologie du Bassin du Rhin (KHR),* Den Haag (in German and in French).

Commission d'Etude des Crues du Rhin (Rapport Final). 1978. 78 p. + ann.

Commission Permanente. 2016. Démonstration de l'efficacité des mesures de rétention des crues du Rhin Supérieur entre Bâle et Worms. Rapport intermédiaire : automne 2016. *Convention du 4 juillet 1969 entre la République française et la République fédérale d'Allemagne sur l'aménagement du Rhin entre Strasbourg/Kehl et Lauterbourg/Neuburgweier,* 24 p (in German and in French).

Convention. 1982. Convention du 6 décembre 1982 modifiant et complétant la Convention additionnelle du 16 juillet 1975 à la Convention du 4 juillet 1969 entre la République Française et la République Fédérale d'Allemagne au sujet de l'aménagement du Rhin entre Strasbourg/Kehl et Lauterbourg/Neuburgweier.

Dister, E., D. Gomer, P. Obrdlik, P. Petermann, and E. Schneider. 1990. Water Management and Ecological Perspectives of the Upper Rhine's Floodplains. *Regulated Rivers. Research and Management* 5: 1–15.

Kuhl, D. 1992. 14 Years of Artificial Grain Feeding in the Rhine Downstream the Barrage Iffezheim. In *Proceedings 5th International Symposium on River Sedimentation,* 1121–1129, Karlsruhe.

Mostert, E. 2009. International Co-operation on Rhine Water Quality 1945–2008: An Example to Follow? *Physics and Chemistry of the Earth* 34: 142–149.

Schmitt, L., and J.N. Beisel. 2015. Les projets de restauration du Rhin Supérieur: vers la mise en place d'un observatoire transfrontalier et transdisciplinaire? In

Final Proceedings International Conference Integrative Sciences and Sustainable Development of Rivers, 3 p, Lyon.

Schmitt, L., D. Roy, M. Trémolières, C. Blum, E. Dister, U. Pfarr, and V. Späth. 2012. 30 Years of Restoration Works on the Two Sides of the Upper Rhine River: Feedback and Future Challenges. In *Final Proceedings International Conference Integrative Sciences and Sustainable Development of Rivers*, 101–103, Lyon.

Schmitt, L., J. Houssier, B. Martin, M. Beiner, G. Skupinski, E. Boës, D. Schwartz, D. Ertlen, J. Argant, A. Gebhardt, N. Schneider, M. Lasserre, G. Trintafillidis, and V. Ollive. 2016. Paléo-dynamique fluviale holocène dans le compartiment sud-occidental du fossé rhénan (France). *42ᵉ supplément à la Revue archéologique de l'Est*, 15–33.

Staentzel, C., F. Arnaud, I. Combroux, L. Schmitt, M. Trémolières, C. Grac, H. Piégay, A. Barillier, V. Chardon, and J.-N. Beisel. Accepted, Under Revision. How Do Instream Flow Increase and Gravel Augmentation Impact Biological Communities in Large Rivers: A Case Study on the Upper Rhine River. *River Research and Applications*.

Trémolières, M., L. Schmitt, S. Defraeye, C. Coli, E. Denny, M. Dillinger, J.C. Dor, P. Gombert, A. Gueidan, M. Lebeau, S. Manné, J.P. Party, P. Perrotey, M. Piquette, U. Roeck, A. Schnitzler, O. Sonnet, J.P. Vacher, V. Vauclin, M. Weiss, N. Zacher, and P. Wilms. 2008. Does Restoration of Flooding and Reconnection of Anastomosing Channels in the Upper Rhine Floodplain Improve Alluvial Ecosystem Functions and Biodiversity? In *Proceedings IVth ECRR International Conference on River Restoration 2008*, ed. B. Gumiero, M. Rinaldi, B. Fokkens, 825–836. European Centre for River Restoration, Centro Italiano per la Riqualificazione Fluviale.

Uehlinger, U., K.M. Wantzen, R.S.E.W. Leuven, and H. Arndt. 2009. The Rhine River Basin. In *Rivers of Europe*, ed. K. Tockner, U. Uehlinger, and C.T. Robinson, 199–246. London: Elsevier.

Wetter, O., C. Pfister, R. Weingartner, J. Luterbacher, T. Reist, and J. Trösch. 2011. The Largest Floods in the High Rhine Basin Since 1268 Assessed from Documentary and Instrumental Evidence. *Hydrological Sciences Journal* 56 (5): 33–758.

Wieriks, K., and A. Schulte-Wulwer-Leidig. 1997. Integrated Water Management for the Rhine River Basin, from Pollution Prevention to Ecosystem Improvement. *Natural Resources Forum* 21 (2): 147–156.

PART II

Urban Streams

CHAPTER 5

Managing Floods in Mediterranean-Climate Urban Catchments: Experiences in the San Francisco Bay Area (California, USA) and the Tagus Estuary (Portugal)

Pedro Pinto, Raymond Wong, Jack Curley, Ralph Johnson, Liang Xu, Len Materman, Mitch Avalon, Graça Saraiva, Anna Serra-Llobet, and G. Mathias Kondolf

Abstract The San Francisco and Lisbon estuaries share many geographical similarities, but their different governance makes for interesting comparisons. Many tributaries to San Francisco Bay were channelized by the US Army Corps in the 1950s–1970s. The design flaws of these projects (such as their having ignored sediment) are manifest as local governments

P. Pinto
University of Lisbon, Lisbon, Portugal

R. Wong
University of California Berkeley, Berkeley, CA, USA

GHD, San Francisco, CA, USA

J. Curley
Marin County Flood Control and Water Conservation District, San Rafael, CA, USA

© The Author(s) 2018
A. Serra-Llobet et al. (eds.), *Managing Flood Risk*,
https://doi.org/10.1007/978-3-319-71673-2_5

93

now struggle to maintain and operate them. Local agencies in the San Francisco region have used a range of innovative tools to solve these flood risk problems, such as creation of a new governance structure encompassing multiple jurisdictions around a stream, adoption of a 50-year plan to convert aging concrete channels into natural creeks, and implementation of projects that combine flood risk reduction with ecosystem and social benefits. Such a multi-purpose project built on the Ribeira das Jardas Stream near Lisbon has proven highly successful, especially as an urban social space.

Keywords San Francisco Bay region • Tagus estuary • US Army Corps of Engineers • Polis program • 50-year plan

R. Johnson
Alameda County Flood Control and Water Conservation District, Hayward, CA, USA

L. Xu
Santa Clara Valley Water District, San Jose, CA, USA

L. Materman
San Francisquito Creek Joint Powers Authority, Menlo Park, CA, USA

M. Avalon
Contra Costa County Flood Control & Water Conservation District, Martinez, CA, USA

G. Saraiva
University of Lisbon, Lisbon, Portugal

A. Serra-Llobet (✉)
University of California Berkeley, Berkeley, CA, USA

Aix-Marseille University, Marseille, France

G. M. Kondolf
University of California Berkeley, Berkeley, CA, USA

University of Lyon, Lyon, France

5.1   INTRODUCTION

Pedro Pinto and Anna Serra-Llobet

5.1.1   Urban Catchments in the San Francisco Bay Area and the Lisbon Estuary Area

The San Francisco Bay Area and Lisbon estuary area share several similarities including climate (Mediterranean climate) and topography, with both regions surrounding large "drowned-valley"-type estuaries. Smaller catchments, draining directly into the estuary, have been heavily built over in both regions, which contributed to increased exposure to urban flooding and further reducing the already short concentration times. The Mediterranean climate is characterized by extreme variability of precipitation patterns, and localized events of heavy rainfall are responsible for sporadic, but severe, flood events.

Despite these similarities, the urbanization of both regions had significant differences. San Francisco Bay (actually an estuary, not a "bay") experienced very fast urban expansion within a very short time frame: from negligible urbanization in as late as the 1850s, the estuary was almost completely encircled by urbanization little over a century later. This development took place in an era of strong reliance on heavy flood control infrastructure. Most creeks in the region were confined to concrete canals and their former floodplains extensively built over. In contrast, urban expansion in Lisbon was a slow, multi-secular process and, unlike San Francisco, was only characterized by rapid suburban expansion in the last three to four decades. As such, large portions of the former countryside were left undeveloped until a time when modern environmental protection standards were beginning to be seriously considered. Although a few urban areas in the region were expanded so as to encroach on floodplains, most of the urban expansion took place over the surrounding hills and plateaus instead (Pinto and Kondolf 2016).

Perhaps more striking are the disparities in the way flood control/management is conducted in both regions. Portugal is a unitary state, with national government being directly responsible for most issues related with environmental standards, flood regulation, and, in most cases, even the conception and building of flood defense infrastructure. Municipalities have the local planning mandate and have great autonomy in terms of zoning decisions, but have almost continuously been able to uphold a relatively strong public trust over stream banks. Except for a short period starting in the late 1970s and extending to the early 1990s, local plans have been reasonably effective in protecting floodplains from urban development, and it is the municipalities that are tasked with the day-to-day management of stream banks, including bank maintenance. However, they often lack the expertise and money and defer to the water agency, now a part of the Portuguese Environmental Agency, in the design and implementation of more complex flood detention or flood mitigation initiatives. Besides flood management, the agency is also responsible for wastewater regulation, water permitting, environmental protection, and habitat restoration. The agency also oversees the river basin management plans, including the one for the Tagus Basin, theoretically allowing very strong synergies in the coordination and management of these different issues.

In contrast, the San Francisco metropolitan region experienced rapid urban expansion, especially post-World War II, in a context of weak regional and local land-use regulations. Also, city governments had a very strong planning mandate but often not in direct articulation with flood control districts, which nowadays mostly share the same boundaries as counties (Fig. 5.1). These districts are in charge of flood management (stormwater and runoff) in urban areas. In general, they only do flood control, as wastewater treatment, water supply, water conservation, and so on are run by other districts with different boundaries. Typically, these flood control districts are reasonably well funded, especially when compared with the limited resources of the Portuguese Environmental Agency but, in contrast, are constrained by narrow mandates, which have discouraged the integration of local land-use solutions with flood management. Although recent budget constraints have virtually stalled large-scale public works, Portuguese agencies and local governments work within a more favorable legal framework, and the possibility of integrating flood defense with, for example, urban redevelopment or habitat restoration appears to be much facilitated.

In this chapter, Raymond Wong and Matt Kondolf review the many flood control projects built in tributaries to San Francisco Bay during the building boom of the 1950s, 1960s, and 1970s by the US Army Corps of Engineers (USACE) and turned over to local governments to operate. Unfortunately these projects suffered design flaws, such as ignoring the effects of the commonly high loads of sediment, whose accumulations posed unrealistic maintenance burdens on local governments. Jack Curley reviews the seven-decade-long history of flood control efforts on Corte Madera Creek in Marin County. Ralph Johnson describes the legacy of

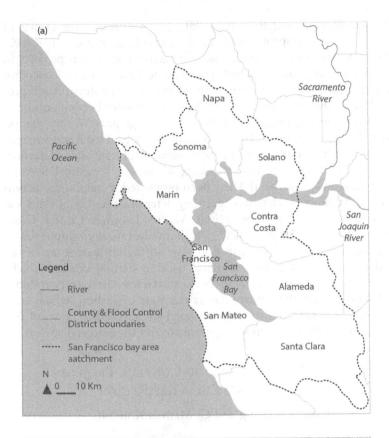

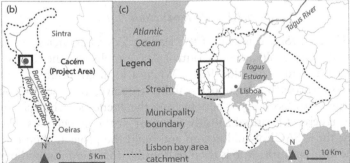

Fig. 5.1 San Francisco Bay Area catchment boundary in California (United States) (**a**). The Tagus estuary catchment boundary in Portugal and location of Ribeira das Jardas Project (**b–c**)

poorly sited development from the 1950s and 1960s and the efforts of Alameda County to improve predictions of flood-prone areas. Liang Xu illustrates the evolution of flood risk management with two projects in Santa Clara County, one of which (Guadalupe River) evolved over five decades into a multi-purpose project to provide flood protection and preserve habitats. San Francisquito Creek forms the border between multiple cities and counties, and thus was difficult to manage prior to the creation of a Joint Powers Authority (JPA) involving all relevant local government agencies, a creative solution to the governance problem. Len Materman reports on the JPA's challenges working within the lengthy US Army Corps process and how the JPA has been motivated to take some action on its own. As concrete channels age, they deteriorate and confront local agencies with difficult choices. Mitch Avalon describes Contra Costa County's innovative "50-year plan", which involves local communities to start the planning process now, to find ways so that at least some of the concretized channels can be converted back to natural creeks. Finally, switching estuaries to the Tagus in Portugal, Graça Saraiva reports on a successful project on the Ribeira das Jardas west of Lisbon, which converted a concrete culvert back into an open stream, reducing flood risk while creating a vibrant urban social space around the restored stream.

5.2   A History of Flood Control Projects in the San Francisco Bay Region

Raymond Wong and G. Mathias Kondolf

5.2.1   Introduction

In the USA, the USACE is one of the leading federal agencies on flood management. The 1936 Flood Control Act established the USACE's flood control mission (Arnold 1988). When a local community experiences flooding problems beyond its ability to solve, the local agencies could partner with USACE, to benefit from the agency's technical expertise and to receive a financial subsidy for most of the project cost. Most commonly, the USACE would design and build the project, then turn it over to the local sponsor, who is then responsible for operation and maintenance (O&M) (Carter and Stern 2010). The 1986 Water Resources Development Act (WRDA) revised the "cost-sharing" policy, so that federal government covered a much smaller percentage of the total project cost than had previously been the case.

Fig. 5.2 Grayson Creek channel right after completion in 1957 (Courtesy of Contra Costa County Flood Control and Water Conservation District)

Historically, USACE flood management has been heavily skewed toward conventional structural approaches. Most projects were designed to provide predictable flow conveyance capacity in the smallest possible footprint. Ecological values were mostly ignored. In addition, many of these project designs were based on unrealistic assumptions; notably, they did not adequately account for sedimentation, resulting in massive sedimentation problems in the flood control channels (Williams 1990).

As a result, many local agencies cannot afford the significant O&M requirements they inherited with the projects. Many of these projects are plagued by chronic problems on sedimentation and inadequate level of flood protection. It presents a challenge to the local agencies on how to provide adequate level of services for flood protection, while balancing project life cycle cost.

In the San Francisco (SF) Bay region, many flood control projects designed and built by the USACE during the 1960s and 1970s included reaches of narrow concrete channels (Fig. 5.2). Some of these projects are now recognized as undersized and pose significant O&M challenges due to sedimentation. As these O&M problems emerged, they have been treated as independent problems unique to the individual projects. However, in reality these projects share commonalities in planning and design approaches, and in their resulting O&M problems (Samet 2007). Wong (2014) analyzed nine such projects in the SF Bay region and found that the current channel sediment removal cost is about five times higher than the original estimate of the project design, adjusted to present value. Five of the nine projects have existing channel capacity data, but none of them has 100-year flow capacity due to a combination of watershed urbanization and channel sedimentation. Notably, six of the nine projects relied on "land enhancement benefit", defined as the net incomes and property values of turning undeveloped floodplain into urban development, as a benefit to produce a positive cost benefit ratio. If the land enhancement benefit is removed from the cost benefit ratio, three of the six projects would have had cost benefit ratio below 1, meaning the projects would not be economically justifiable to proceed.

As an illustration of the challenges faced by local agencies as they attempt to maintain flood control function, we describe two case studies: San Lorenzo Creek and Walnut Creek.

5.2.2   San Lorenzo Creek, Alameda County

San Lorenzo Creek drains a 124 km² (48 mi²) catchment, flowing westward from Cull Canyon, Crow Canyon, and Palomares sub-catchments into SF Bay. To alleviate recurring flooding in the downstream floodplain, the 1954 Flood Control Act authorized San Lorenzo Creek Flood Control Project. The project, completed in 1962, included 8.4 km (5.2 mi) of concrete channels and earthen trapezoidal channel flanked by levees. The construction cost was $4.28 million at the time (equivalent to about $60 million in 2010 dollars), with a calculated 1.17 benefit cost ratio.

After the project was completed, urban development increased the 100-year peak flow from 227 m³/s (8016 ft³/s) to 468 m³/s (16,527 ft³/s). Since the project design flow is 275 m³/s (97,011 ft³/s), the project no longer has 100-year flow capacity.

Unrelated to the flood control project, in the 1960s the county constructed reservoirs on two tributaries, Cull Canyon and Palomares, funded by the Davis Grunsky Act, to provide recreation and water supply benefits. The reservoirs also provided some ancillary flood storage, but their capacities declined rapidly from sedimentation, both down to only 20% of their original capacities by 2003. Sedimentation in these reservoirs has reduced sediment delivery to the downstream flood control channel, although the third major fork of San Lorenzo Creek, Crow Canyon, still supplies sediment to the downstream channel without impairment.

The downstream reaches, within the engineered flood control channel, have relatively flat slopes and thus are natural sites for sedimentation. In the past, flood overflows would distribute the sediment load over the marsh plain, but with channel constriction and levees, the sediment is either carried into the bay or deposited in the channel. Since 1962, the county has spent $4.1 million (2010 dollars) on sediment removal from the flood control channel. If the total cost were to include the estimated cost to remove all sediment in both reservoirs, it would increase to $52 million (2010 dollars). It is an order of magnitude higher than the total O&M cost estimate in 1954, at $3.5 million (2010 dollars).

5.2.3   Case Study: Lower Walnut Creek, Contra Costa County

Walnut Creek drains a 378 km² (146 mi²) watershed, from the headwaters of its Pine Creek tributary near the summit of Mount Diablo at 1173 m (3849 ft) above mean sea level, flowing northward and dropping to sea level at San Pablo Bay. The mean annual precipitation is 530 mm (21 in) (CCCFCWCD 2003). The Contra Costa County Flood Control and Water Conservation District (Contra Costa County) is responsible for Walnut Creek watershed planning and flood management. Repeated floods in the mid-twentieth century prompted the Lower Walnut Creek Flood Control project with USACE, authorized under the Flood Control Act in 1960 (USACE 1963). The project constructed in 1965 included 22.7 km (14.1 mi) of earth and concrete channel and levee sections, at the sections of Walnut Creek between Rudgear Road and the outfall at Suisun Bay. The project cost was $31,500,000 in 1964 dollars, with 1.3 cost benefit ratio (USACE 1964).

The project was designed for a Standard Project Flood of 708 cms (25,000 cfs) at the bay. However, a 2008 reevaluation of the project by the USACE estimated the 100-year design flow as 884 cms (31,200 cfs), but channel flow capacity to be only 566 cms (20,000 cfs) (RDG 2013). Thus, the Lower Walnut Creek Flood Control project not only cannot convey the 100-year flow, but also did not maintain its original design capacity.

The Lower Walnut Creek Flood Control project was designed with a flat bottom and no low-flow channel. The project has been plagued with sediment issues since its construction in 1965. In 1972, USACE revised its estimate for sediment deposition in the flood control channel upward from that stated in the project General Design Memorandum, from

28,000 m³/yr (36,000 yd³/yr), out of an estimated sediment load of 138,000 m³/yr (180,000 yd³/yr), a 20% trap rate. The revision estimated the flood control channel would trap 122,000 m³/yr (160,000 yd³/yr) out of 192,000 m³/yr (250,000 yd³/yr) sediment supply, a 65% trap rate, and a more than fourfold increase in predicted sedimentation (USACE 1972).

Between 1973 and 1989, USACE and the Contra Costa County removed approximately 861,000 m³ (1,126,000 yd³) of sediment in the creek. In the early 1990s, the district estimated that 497,000 m³ (650,000 yd³) of sediment had accumulated in the area dredged by USACE in 1973. After significant efforts to secure regulatory permits for sediment removal, Contra Costa County concluded that the dredging work was unlikely to be permitted due to significant environmental impacts and that mitigation costs would far exceed the county's financial resources.

In 2007, the USACE released a nationwide evaluation of flood control systems and included Lower Walnut Creek in the deficient category. As a result, Contra Costa County implemented the Interim Protection Measures Project and removed 153,000 m³ (200,000 yd³) of sediment between BNSF Railroad and Clayton Valley Drain. Contra Costa County continues to evaluate options to sustainably maintain the lowest 4 km (2.5 mi) of Lower Walnut Creek, to balance ecological function and flood protection benefits, and meet the USACE maintenance requirements. The county concluded that there where significant permitting hurdles to continue the needed dredging operation at the lower reach. In addition, since the lower reach is away from the urban areas, the residual risk is relatively low. At the Contra Costa County's request, the lowest 4 km (2.5 mi) reach of the creek was de-authorized from the USACE Lower Walnut Creek flood control project (U.S. Congress 2014). As a result, Contra Costa County can redesign the lower reach with a different design frequency, as long as it does not impact the capacity of the upstream project.

5.2.4   Conclusion

San Lorenzo and Walnut Creek illustrate key attributes of USACE flood control projects in the region as borne out by Wong's (2014) study of nine such projects:

- Cost Benefit Analysis: A key USACE project planning tool, the benefit cost ratio for a project must exceed unity or the project is rejected.

This approach creates an incentive to select the lowest capital cost alternative to provide flood protection benefits, often at the expense of environmental and social values, as well as unrealistic and underestimated O&M requirements. Moreover, most of the projects built in the SF Bay region in the 1960s and 1970s relied on enhanced land values (which is to say, the anticipated value of houses built in the floodplain was induced by the flood protection promised by the project) to yield a positive benefit cost ratio.

- Project Design: The clear water and supercritical flow assumptions used in the designs implicitly assume that sediment does not affect flow hydraulics, which is manifestly incorrect. It was a fundamental design flaw in all studied flood control projects (Wong 2014). The basic geomorphic principle that sediment deposition would be expected on distal alluvial fan reaches was simply ignored. Thus, the sedimentation rate was underestimated, and as a result, channel capacity was overestimated.

- Operation and Maintenance: Federal appropriation only covers the capital project cost. Local sponsors must fund the O&M, but many projects were not designed for efficient maintenance, nor were maintenance requirements properly estimated in the original project documents. Inadequate O&M reduces project performance, a significant short fall whose consequences have fallen on the local sponsors, who have inherited projects they cannot afford to maintain. Consequently, project performance deteriorates over time, resulting in a false sense of security for residents "protected" by these projects.

5.3   CORTE MADERA CREEK IN MARIN COUNTY

Jack Curley

5.3.1   Introduction: The Ross Valley Catchment

The Ross Valley Catchment is located approximately 32 km (20 miles) north of San Francisco in Marin County. The Ross Valley contains over 70 linear km (44 mi) of stream channels and covers an area of approximately 72.5 km² (28 mi²) (MCFCD 2017a). The catchment starts in the hills above the Town of Fairfax, where the stream is called Fairfax Creek until the confluence of the tributary San Anselmo Creek coming out of the northwest hills. It then becomes San Anselmo Creek after a major confluence in downtown San Anselmo where it is joined by Sleepy Hollow Creek and Sorich Creek. It flows into the Town of Ross and is called Corte Madera Creek after the confluence with Ross Creek and then to San Francisco Bay at City of Larkspur.

The Ross Valley Catchment is one of eight flood risk management zones managed by the Marin County Flood Control and Water Conservation District (the District). The District was created in 1953 and its geographical boundaries coincide with those of Marin County. This district, under the authority of the Marin County Board of Supervisors, is responsible for enacting measures that reduce the risk of flooding (MCFCD 2017b).

5.3.2   History of Flooding

The Corte Madera Creek catchment in Marin County, known as Ross Valley, has a long history of large floods. Damaging floods occurred in the catchment in calendar years 1914, 1925, 1937, and 1942. Since 1951, when a USGS gauge was installed on Corte Madera Creek in the Town of Ross, flood flows have been recorded in calendar years 1951, 1952, 1955, 1958, 1967, 1969, 1970, 1982, 1983, 1986, 1993, and 2005. Of these, the two most severe floods occurred in 1982 and 2005, with peak discharges of approximately 200 m^3/s (7200 ft^3/s) and 190 m^3/s (6800 ft^3/s) respectively, the percent-annual-chances (i.e., probabilities) of which were approximately 0.6% and 1%, respectively. The 1955 flood was an approximate 4-percent-annual-chance flood (Stetson Engineers 2011).

Residents of the valley have been working to find a solution to the chronic flooding since early in the twentieth century. In January 1934, the Kentfield Chamber of Commerce called on federal officials to help fix flooding problems along Corte Madera Creek. They warned that the damage caused by the 1925 flood could not be allowed to happen again. A survey of Corte Madera Creek was authorized under the Flood Control Act of December 1944, and the USACE completed a preliminary examination report in 1946. Another round of serious flooding in the early 1950s amplified the calls for action by the local communities. After the December 1955 flood, Ross Valley residents stormed the county supervisors' chambers, demanding formation of a flood control entity for Ross Valley in April 1956. Finally, in the Flood Control Act of 1962, the US Congress authorized the design and construction of the Corte Madera Creek Flood Control Project.

5.3.3   Corte Madera Creek Flood Control Project

The estimated federal cost share for the project was 97% or $5,534,000 (in 1960s' dollars) with local interests contributing $158,000. The project was conceived to consist of six units with a concrete-lined channel extending approximately 6.5 miles from the SF Bay upstream into the Town of Fairfax. It was designed to convey approximately 215 m³/s (7600 ft³/s) or a 250-year flood event. Following two flood events in 1962 and 1963, Congress amended the project under the Flood Control Act of 1966 to reduce the local cash contribution from 3% to 1.5%. In 1968, the USACE completed Units 1 and 2. Together they consisted of a dredged, earthen trapezoidal channel extending 4.8 km (3 mi) from SF Bay to Bon Air Road in Kentfield, a 365 m (1200 ft) settling basin and a concrete-lined channel in the upper 457 m (1500 ft) of Unit 2. After another flood in 1969, work continued on the 1066-m (3500 ft)-long Unit 3, creating a combined 1524 m (5000 ft) of concrete channel, terminating 182 m (600 ft) downstream of the Lagunitas Road Bridge in the Town of Ross.

With Units 1, 2, and 3 constructed, work was halted in 1974 after the settlement of the litigation brought by the Town of Ross and by growing environmental concerns of property owners whose residences/businesses were directly adjacent to the creek. Efforts to find a preferred local solution continued but local support diminished. After the highest flood of record occurred in 1982 and other large floods in 1983 and 1986, Congress authorized the project again in the WRDA of 1986, still limiting the extents of the project to the upstream end of the creek in the Town of Ross, half the distance to the upstream end of the catchment.

After the flood of 1982, local concern grew about the efficacy of the existing project design and was reflected in comments to the supplemental environmental impact documents. During the 1982 event, a storm that delivered 30 cm (12 in) of rain in 32 hours to the valley, local people saw that the concrete channel was unable to pass the peak flow of 203 m³/s (7200 ft³/s), which was less than the design flood peak, that is, the 250-year flood event with a peak flow of 220 m³/s (7800 ft³/s). Damages were estimated at $80 million in the Ross Valley with more than 35 homes destroyed.

Due to those concerns and the results of engineering studies by private sector consultants hired by the local sponsor, an extensive sedimentation study was carried out by the USACE Waterways Experiment Station (WES) (now the USACE Engineer Research and Development Center

[ERDC]) in 1989. It determined that the flow capacity of the existing concrete-lined channel was significantly less than the 100-year flood event flow. Discussion continued, and the project was reclassified from active to deferred status pending an endorsement of a new consensus plan by the local sponsor. It was determined that building a project to the 100-year level of protection was environmentally unacceptable, and the community agreed on building to a flow capacity of 152 m³/s (5400 ft³/s), approximately the 25-year flood level of protection. The project was reactivated in 1998, and efforts have continued since then to find an acceptable design for completing Unit 4, that is, the last 182 m (600 ft) of the project. The project is moving forward albeit very slowly due to lack of federal funding.

Unit 4 remains an authorized project. In February 2014, the District signed an agreement with the USACE to share the costs of a feasibility study to reevaluate the authorized project (MCFCD 2017c). Under this agreement, the USACE will contribute 50% and Marin County Flood Control will contribute 50% of the cost.

In December 2015, the environmental review process began with the "Notice of Preparation/Intent" and the "Notice of Scoping Meeting" for a joint Environmental Impact Statement/Report (EIS/EIR) for the project (MCFCD 2017d). The Scoping Meeting was held in February 2016.

In August 2016, the USACE project team presented a list of possible alternatives at a public meeting in the Town of Ross to reduce the risk of flooding. Alternatives include top of bank flood barriers, setback flood barriers, and expanding or widening along sections of the earthen and concrete channels. An alternative for an underground bypass channel is also being considered. Removal of the existing fish ladder at Town of Ross is also included (MCFCD 2017e) (Fig. 5.3). USACE is currently modifying the hydraulic model to fully analyze these alternatives. The preliminary cost estimate for the project is $14 million, but this estimate is subject to change based on the results of the feasibility study and the EIR/EIS (MCFCD 2017d).

On December 31, 2005, a New Year's Eve flood devastated the Ross Valley, renewing calls for flood control measures. Flood waters in downtown San Anselmo were nearly 1.2 m (4 ft) high. Damages were estimated to be above $90 million valley-wide. The town halls of Fairfax and San Anselmo required extensive and lengthy rebuilding. Some long-standing local businesses closed for good. Fairfax Creek topped its banks and water flowed through downtown along the historic creek path now developed into a busy downtown and residential area.

Fig. 5.3 Boundary between Unit 3 and Unit 4 in Ross Creek. Fish ladder separating concrete channel from earthen channel. Unit 4 extends 180 m (600 ft) upstream of this fish ladder (Courtesy of Marin County Flood Control District)

All of the tributaries of the upper half of the catchment come together in downtown San Anselmo at Bridge Street. Topping the banks there, the floodwaters flowed through the streets in San Anselmo, the Town of Ross, through Kentfield, finally returning to the main channel in lower Kentfield. The earthen channel, with its 344,050 m³ (450,000 yd³) of sedimentation, passed this flood, later determined to be a very close approximation of the 100-year flood.

Since the flood of 2005 was contained in the earthen channel, the District is altering its approach to the maintenance of the channel and proposing to dredge to the 100-year level in future years. Under the original agreement, the District is required to dredge to the 250-year level. This must be negotiated with the USACE after the completion of Unit 4 as part of the revision of the Operations and Maintenance Manual. For the very long term, the program looks to increase tidal prism as a means to increase flushing of sediment and as a potential buffer for sea level rise.

The District continues to partner with the San Francisco District of the USACE to complete Unit 4. In 2006, the District launched a catchment-wide flood control program to increase capacity to the 100-year level throughout the catchment. The success of the measures in the upper catchment is dependent on the completion of the federal project which is the furthest downstream project in the catchment. The District has provided updated geometry for the catchment HEC-RAS model to the USACE and a local non-profit, the Friends of Corte Madera Creek, provided state-of-the-art fish passage designs for the Unit 4 reach through a grant from National Fish and Wildlife Foundation. The Flood Control District has provided detailed suggestions to the USACE on a number of measures that would integrate the upstream end of Unit 4 with the local catchment program in the Town of Ross.

Despite its lengthy history, there is optimism that Unit 4 can be completed, and local leaders are working to find a way to help the USACE San Francisco District complete the design and environmental process. In the meantime, the local effort is under way in the rest of the catchment. The Flood Control District recently received a grant from the California Department of Water Resources through the Proposition 1E Stormwater Flood Management program for $7.6 million (50% of the project cost) to retrofit the local water supply reservoir, Phoenix Lake, to function as a detention basin during the 100-year event. The Phoenix Lake project alone could reduce the peak 100-year flow by half, 18.4 m^3/s (650 ft^3/s) of 38 m^3/s (1340 ft^3/s) at Ross gauge, and deliver the flow to Unit 4 that it is expected to be built to handle, that is, 150 m^3/s (5400 ft^3/s).

5.4   EVOLUTION OF FLOOD MANAGEMENT IN ALAMEDA COUNTY

Ralph Johnson

5.4.1   Introduction

Much of Alameda County is in a floodplain. Until the 1950s and 1960s, many parts of the County were subjected to repeated flooding that closed businesses and schools, damaged agricultural crops, interrupted transportation and utility services, destroyed homes, and took lives.

At the request of Alameda County residents, the California State Legislature created the Alameda County Flood Control and Water Conservation District in 1949. Throughout the 1950s and 1960s, cities and unincorporated areas joined the Flood Control District to receive protection from devastating floods. The cities and unincorporated areas were grouped into "zones" which generally corresponded to catchment boundaries. There are now ten individual flood control zones, and since 1949, the Flood Control District has steadily constructed strategic flood control improvements. Each zone or catchment has a "master plan", and projects were constructed generally from downstream to upstream, always following the "master plan".

Even though few in the 1950s could have imagined the magnitude of development that has occurred in Alameda County, the flood control system was constructed assuming full build-out of the land. This system includes pump stations, reservoirs, and hundreds of miles of pipeline and channels. Recent estimates placed the value of the flood control infrastructure at over $850 million. The system components have been paid for with a combination of property taxes, benefit assessments, special federal and state project funding, and developer fees.

5.4.2   The Concept of Wise Use of Floodplains in the 1950s and 1960s

During the 1950s and 1960s we were convinced that our concrete and steel engineering could manipulate the earth as we desired without any adverse impacts. We moved streams, covered creeks, changed shorelines, and replaced floodplains with cities. Our engineering produced a vast network of earthen and concrete channels, underground pipes, and culverts.

Benefit cost ratios were used to justify constructing flood control projects, provided that anticipated benefits exceeded project costs. In this era, the calculated "benefits" included the value of the land made available for new development by protection against flooding (see Sect. 5.2). Removal of the threat of floods also permitted more comprehensive local planning for future development and allowed freeway construction to proceed. As the freeways were part of the National Defense Highway System, major benefits could be included.

So as long as the risk to life and property could be mitigated at a reasonable cost, then the development of the floodplain was viewed as "wise". If locating development in the floodplain disrupted the natural functions of the floodplain, this was not viewed as a problem, probably because it was thought that there were always other natural floodplains left.

The work of the Flood Control District is far from complete. Aging facilities must be replaced and (as noted in Sect. 5.2) efforts to keep flood control channels clear of silt and debris never cease. Increased focus on

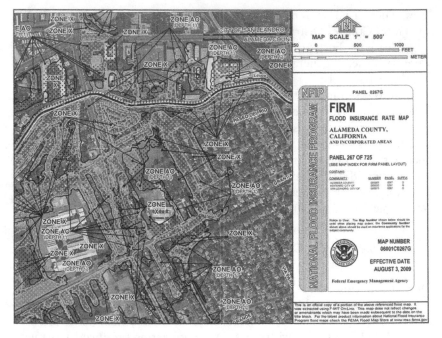

Fig. 5.4 The 2009 Flood Insurance Rate Map of northwest San Lorenzo, California, showing San Lorenzo Creek, Zone AO (the 100-year regulatory floodplain) and Zone X (other areas considered moderate or low risk) (Source: US Federal Emergency Management Agency)

controlling stormwater pollution and restoring stream habitat is the new challenge for the District.

5.4.3   Appeal of FEMA Floodplain Mapping in 1999

A FEMA remapping project completed in 1999 placed over 19,000 new parcels in Alameda County within a 100-year floodplain (Fig. 5.4). Homeowners in Alameda County were faced with paying millions of dollars of flood insurance. Additionally, the Flood Control District was faced with paying for a whole new set of flood capital improvement projects.

Using a new sophisticated hydrological computer model, District engineers more accurately predicted how much rain would fall in a major storm and how much would run off into the District's creeks. Then, they evaluated the flood water levels caused by this runoff. Finally, they used their own digital mapping and extensive knowledge of local hydraulics to map the probable extent of flooding.

As a result, the District was able to successfully remove over 12,000 of the 19,000 parcels from the floodplain. The Alameda County Flood Control District's advocacy on behalf of these homeowners means that they do not have to buy expensive flood insurance, which typically costs over $1000 per year. This was the first time a local agency has so thoroughly and successfully appealed FEMA's flood studies. The technical quality of the District's reanalysis was so high that FEMA initiated a new program that allows local jurisdictions to oversee future floodplain analyses.

5.4.4   *Strategic Plan for the Twenty-First Century*

Using the experience gained during the 1999 appeal, the Flood Control District has updated floodplain mapping throughout the County to better inform County residents of the risks they have of living in a floodplain and to guide a maintenance and capital construction program that works toward minimizing the risks of living in a floodplain. This work has influenced the District to adopt a strategic plan for use of floodplains for the twenty-first century that includes sustainable Flood Control management, responsible environmental restoration, and clean water collaboration.

5.5   FLOOD PROTECTION PROJECTS IN SANTA CLARA VALLEY

Liang Xu

5.5.1   *Introduction*

Flood protection projects in urban streams must balance land use, right-of-way limitations, and preservation or enhancement of existing riparian habitat corridors while providing protection against the 100-year flood. This section describes two large projects in Santa Clara Valley, the Guadalupe River in the city of San Jose, and Permanente Creek in Mountain View. Both projects demonstrate how the Santa Clara Valley Water District (SCVWD) has dealt with the challenges of providing flood protection and generating community support. SCVWD is the largest multi-purpose special district in California, providing water supply, flood management, and catchment stewardship to a population of 1.9 million, including Silicon Valley.

5.5.2   The Guadalupe River Park and Flood Protection Project

The Guadalupe River drains 360 km² (140 mi²), flowing north through Santa Clara Valley and the center of San Jose to debouch into the southern end of the San Francisco estuary. The Guadalupe River Park and Flood Protection Project (or Downtown Guadalupe River project) extends from Interstate 880 to Interstate 280 in the city of San Jose. This $350 million multi-purpose project was completed in 2005 to provide flood protection to the city's technology and commercial industries and established residential neighborhoods; protect and improve the water quality of the river; preserve and enhance the river's habitat, fish, and wildlife; and provide recreational and open space benefits. The SCVWD served as the local sponsor, working with the USACE.

Persistent flooding problems initially led to calls for a river improvement project in the early 1960s, but for three decades the city of San Jose adopted a vision for the river based on the San Antonio River Walk, a well-known 1.2-km (0.74 mi) reach of the San Antonio River in San Antonio, Texas, along which a paved walkway, cafes, restaurants, and hotels are located. Early proposals included damming the Guadalupe River in a series of lakes to create stable, perennial water (despite the seasonal runoff in the Mediterranean climate), but the dams would have blocked migration of steelhead trout (Kondolf et al. 2013).

The project evolved considerably over the years, with modifications to the project design including channel widening, bridge replacement, and incorporation of a river walk, maintenance roads, and recreation elements, along with extensive planting for environmental mitigation (Fig. 5.5). The listing of two endangered species required added environmental study. The history of the project is long and complex and will not be recounted here, but many modifications and mitigations were undertaken in response to litigation brought by environmental groups and Guadalupe-Coyote Resource Conservation District (Roos-Collins 2007). The modified project included an underground bypass box culvert to carry flood flows around important environmental resources in the natural channel, streambed erosion protection features, terraces, and environmental mitigation to enhance habitat for steelhead trout (*Oncorhynchus mykiss*) and Chinook salmon (*Oncorhynchus tshawytscha*) (required by the Endangered Species

Fig. 5.5 Guadalupe River flood control project looking upstream at outlet of Woz Way Bypass, downtown San José, California (Courtesy of Marin County Flood Control District, 2006)

and Clean Water Acts), with the goal of maintaining cooler water temperatures.

Public access was also incorporated into the project, as the Guadalupe River Park, a 3-mile ribbon of parkland running along the banks of the Guadalupe River in the heart of downtown San Jose, was a resource of regional importance to the people of Santa Clara County and the SF Bay region. In 2009, SCVWD began to work on the Upper Guadalupe River project to provide flood protection for an additional 10 km upstream. Key components of this project have been built, but the project is not yet complete.

5.5.3   Permanente Creek Flood Protection Project

Another project on Permanente Creek is part of the voter-approved Clean Safe Creeks Program. The objectives for the project are to provide flood protection to homes downstream of El Camino Real. The uniqueness of the project is to have multiple offstream detention basins using the city and county parkland to reduce the peak flow. By using detention basins, we can maintain existing right-of-way in urban areas and reduce costs and impacts to communities along the creek (SCVWD 2008).

After reviewing the feasible alternatives using Natural Flood Protection objectives, engagement with the community, and feedback received from citizens, the Permanente Task Force, City staff, and elected officials, the District staff has identified the following elements for the project. This alternative is composed of the following project elements:

- Offstream flood detention facilities in Rancho San Antonio Park and McKelvey Park
- Bypass channel along Hale Creek
- Channel widening along reaches of Permanente Creek and Hale Creek
- Floodwalls north of Highway 101 on levee channels

There will be an opportunity for restoration, habitat enhancement, and trail extension upstream of Highway 101 when designing the project. The project is currently under construction.

5.6   SAN FRANCISQUITO CREEK IN SAN MATEO COUNTY

Len Materman

5.6.1   Introduction

San Francisquito Creek is located approximately (48 km) 30 miles south of San Francisco in the heart of Silicon Valley. It drains a catchment of approximately 116 km^2 (45 mi^2), flowing eastward from the crest of the Santa Cruz Mountains down to the San Francisco estuary. In its course, it is crossed by major highways, rail lines, and infrastructure corridors. The largest landowner in the catchment is Stanford University, which owns much of the headwaters, including a nineteenth-century dam now filled with sediment. San Francisquito Creek itself is about 20 km long (and forms the boundary between San Mateo and Santa Clara Counties). With its tributaries, the stream system provides over 100 km (60 mi) of channels, the last relatively unaltered creek system in the southern part of SF Bay. This creek's fluvial floodplain overlaps with the SF Bay tidal floodplain: of the approximately 8500 parcels in both floodplains, about 42% are in the fluvial floodplain only, about 32% in the tidal floodplain only, and about 26% in both simultaneously. The 14-mile main stem of San Francisquito Creek forms the boundary between San Mateo and Santa

Clara Counties in the lower catchment, where its floodplain extends almost 8 km (5 mi) from northern Menlo Park to south Palo Alto and about 4 km (2.5 mi) from the Bay on the east to Middlefield Road in the west.

The 1998 flood (the largest flow since measurements began in 1930, an approximately 80-year flood) caused $28 million in documented damage to over 1700 properties in Palo Alto, East Palo Alto, and Menlo Park, and closed the major freeway between San Francisco and Silicon Valley. Other recorded flooding and high flow events have occurred in 1955, 1958, 1982, 2002, 2005, 2012, and 2017. The USACE has estimated that damages from a 100-year or 1% flow event on San Francisquito Creek would cause 25 times the financial damages experienced in 1998 and would pose a far greater threat to lives, property, and regional commerce. Following substantial planning efforts in previous decades by government agencies and stakeholders, the 1998 event compelled five Silicon Valley jurisdictions the following year to form a new regional government agency named for the physical features that unites and divides them, the San Francisquito Creek Joint Powers Authority (SFCJPA).

The San Francisquito Creek catchment encompasses the cities of East Palo Alto, Palo Alto, and Menlo Park, among others, and thus the SFCJPA was founded by these three cities, as well as the two countywide flood protection entities on both sides of the waterway: the SCVWD and San Mateo County Flood Control District. The SFCJPA Board of Directors is composed of an elected official from each of these jurisdictions—in 2017, it is the mayor or vice mayor of each city, a county supervisor, and water district director. In addition to its work on the creek, the SFCJPA is addressing flooding, ecosystem, and trail opportunities along the shoreline of SF Bay with its project known as the Strategy to Advance Flood protection, Ecosystems, and Recreation along the Bay (or SAFER Bay), the largest multi-county effort to protect against sea level rise in California in an area with internationally significant assets at risk of tidal flooding today.

In the past five years, the SFCJPA has secured over $83 million in commitments for its projects from its local partners, state and federal governments, and the private sector. The small SFCJPA staff advances its multi-jurisdictional, multi-benefit capital projects with the help of consultant teams and the expertise of staff from the five agencies that established it. While this creek is a boundary between these jurisdictions, it is also what unifies them, and the framework of the SFCJPA has provided a forum for these communities to pursue other activities that cross their borders.

5.6.2   San Francisquito Creek Flood Damage Reduction and Ecosystem Restoration Project

In 2002, the SFCJPA initiated an effort to plan the implementation of its first flood protection project. After cataloging project concepts proposed since the 1950s by the USACE and local agencies, the SFCJPA developed a project and sought formal federal participation by the USACE through a federal catchment-wide Feasibility Study, with the SFCJPA as the local sponsor. In 2005, the SFCJPA Board of Directors elected to put the local project on hold to concentrate on working with the USACE.

The annual appropriations bill or a USACE Work Plan provides the federal government's 50% share of funding for Feasibility Studies. As is the case with many such studies, in the years since the San Francisquito Study began, federal funding has been inconsistent and inadequate to meet the schedule originally envisioned by the USACE. At various times since 2005, the SFCJPA has taken on significant portions of the study's technical analysis with the USACE focused on reviewing conformance with federal standards, and the SFCJPA has provided local funds to the federal study ahead of federal appropriations.

Another area where local resources advanced the overall effort has been in the planning and design of a so-called early implementation project in the furthest downstream reach of the catchment from SF Bay to US Highway 101. In this area, where there is substantial overlap between fluvial and tidal floodplains, the poorest community on the San Francisco Peninsula—East Palo Alto—lies below sea level and is "protected" from creek and tidal flooding by an uncertified levee that has seeped water during recent high flow events (Fig. 5.6). This is perhaps that part of the San Francisco Bay most vulnerable to sea level rise.

In the summer of 2009, a consultant to the SFCJPA produced an analysis of project alternatives in this reach between SF Bay and Highway 101, the most downstream area that would need to be constructed first. That August, the SFCJPA combined this analysis with other studies, including some done by the USACE, and formally requested federal credit for local resources spent on this aspect of the overall federal plan. In 2010, other SFCJPA consultants began the process to complete plans and specifications on this project and to complete the environmental reviews necessary

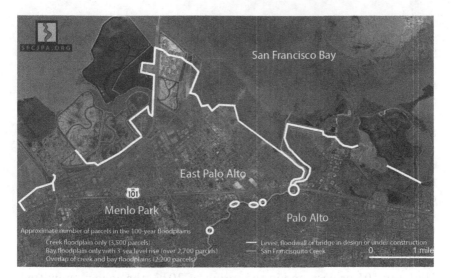

Fig. 5.6 San Francisquito Creek Area projects and floodplains (Courtesy of San Francisquito Creek Joint Powers Authority, 2017)

to begin construction. In the late spring of 2011, the USACE notified the SFCJPA that its 20-month-old application for credit could no longer be entertained because a new crediting policy was being put into effect.

In 2013, given the SFCJPA's progress to complete design and secure local and state funding for this project, the agency had a decision to make regarding whether its SF Bay-Highway 101 project should remain part of the USACE Feasibility Study. While the SFCJPA celebrated having completed environmental documents and submitted its permit applications, it had a dilemma:

- Should it remove the SF Bay-Highway 101 project from the USACE Feasibility Study and ask the USACE to complete a smaller study more quickly so that the SFCJPA could begin construction in the critical area closest to the Bay and then apply to receive credit later, and so that the existing project conditions are maintained.
- Should the SFCJPA delay construction in the Bay-Highway 101 reach and thus jeopardize state funding for construction and keep that project within the USACE Study, and should the SFCJPA ask the USACE to include tidal flooding in the study in order to increase

its benefit-to-cost ratio (BCR) and its chances of receiving federal funding—all of which would add time and cost to the federal study.

In 2012, the USACE Headquarters produced new guidance aimed at reducing the number of—and time to complete—federal Feasibility Studies. In this new guidance, USACE District offices were directed that current and future studies could be completed within three years and for $3 million. To ensure that the San Francisquito Study met these criteria, the SFCJPA was asked to fund a two-day charrette to set the future course of the seven-year long effort. This charrette confirmed the agreement between the USACE and SFCJPA that they would pursue a reduced scope of the federal project by eliminating the Bay-Highway 101 project from the study, a result that matched the SFCJPA's decision to—13 years after the flood of record—no longer wait to fulfill its responsibilities to provide flood protection.

In late 2012, the SFCJPA Board certified the EIR for the Bay-Highway 101 project, and in the spring of 2013, the agency submitted applications for state and federal environmental regulatory permits to begin construction. The permitting process took 35 months, largely due to difficult discussions with the Regional Water Quality Control Board (a state agency). This protracted permitting process increased the project's environmental enhancements but did not change its fundamental features, alignments, and benefits, which include protecting against a 100-year creek flow and 60 cm (2 ft) of freeboard during an extreme tide after 90 cm (3 ft) of sea level rise, which, in total, is about 2.7 m (9 ft) above the higher of daily high tides. This project will also create new marsh habitat from golf course lands and new trail connections, recreate a natural mouth in the adjacent National Wildlife Refuge to this alluvial system for the first time since it was channelized over 75 years ago, and replace a gas pipeline adjacent to East Palo Alto homes dating back to 1931. In February 2016, the SFCJPA's construction manager of the levee and floodwall portion of the project, the SCVWD, put the contract to construct those features out to bid. Construction of those features, as well as enhancements of major utilities led by the SFCJPA, began in the summer of 2016, with work scheduled for completion in late 2018.

Upstream of this reach, the California Department of Transportation (Caltrans) is designing a new Highway 101 bridge over the creek to the

SFCJPA's creek flow specifications—100-year creek flow during an extreme tide and 90 cm (3 ft) of sea level rise at a location that is already influenced by the daily high tide. It is not until the SFCJPA modifies two more bridges and widens the channel in three locations further upstream that the areas upstream (or west) of Highway 101 can be protected against a repeat of the largest flow ever seen, the 80-year flood of 1998. The SFCJPA and its member agencies are in the process of designing these improvements and planning additional work such as bank stabilization and revegetation of riparian habitat, fish passage barrier removal/modification, bike and pedestrian trails, and interpretive signage. And in conjunction with a Corps of Engineers' Environmental Impact Statement and Feasibility Report, the SFCJPA is developing an Environmental EIR to analyze these improvements and alternatives to them that either detain a storm's peak flow in an upstream basin adjacent to the channel on Stanford land, redirect high flows around the floodplain by way of an underground bypass culvert, or contain more water within the channel through new floodwalls. This EIR will tie together the entire project and complement the work of UC Berkeley graduate students who, in the spring of 2012, developed innovative ideas for making SFCJPA projects more sustainable and these communities more livable. Achieving the SFCJPA's stated goal of 100-year flood protection upstream of Highway 101 to remove properties from the flood insurance requirement will require a combination of these four alternatives.

Each of these alternatives comes with difficult political, right-of-way, environmental, and financial issues. No matter which option is chosen by the SFCJPA, the agency believes that the USACE would find a BCR that would support federal investment in this work. A USACE preliminary analysis of economic damages found that there is a likely federal interest in a fluvial-only project or a project that includes both creek and coastal elements, but the BCR is higher with coastal elements.

The SFCJPA has taken a leadership role in the Bay Area by moving forward with a multi-jurisdictional, multi-benefit project that adapts to rising sea level. Beginning in 2012, the agency received grants from state, local, and federal sources, and then funding from Facebook, whose headquarters is surrounded by marsh, to design and complete an EIR for a new 17.7-km (11 miles) Bay coastal system from San Francisquito Creek north to the Redwood City border and from the creek south to the Mountain View border. This project, known as the Strategy to

Advance Flood protection, Ecosystems, and Recreation Along the Bay (SAFER Bay), provides protection against a 100-year tide with 60 cm (2 ft) of freeboard after 90 cm (3 ft) of sea level rise. Protection afforded by the SAFER Bay project will reduce the risk of flooding from rising tides to assets ranging from the headquarters of both Facebook and Google to 19.3 km (12 miles) of state highway, including the main artery between San Francisco and Silicon Valley, to a regional water treatment plant, electrical substation, airport, and postal service facility. SAFER's new infrastructure will enable the restoration of over 400 ha (1000 acre) of marshland by opening these lands to tides along SF Bay and can enhance the heavily trafficked San Francisco Bay Trail.

The SFCJPA believes that strength of the SAFER Bay project lies in the number of assets protected or enhanced, which has brought on multiple partners and multiple benefits for those partners. This extends to construction funding as well, for the diversity of assets protected should dictate the diversity of funding sources paying for protection. Thus, like its funding for planning and design, the SFCJPA has a multi-pronged strategy to fund SAFER Bay's construction, which is likely to exceed $150 million. These sources include traditional opportunities for public works funding like the State of California and federal government, and a new special tax or assessment district passed by properties that would be protected and see flood insurance premiums reduced. But the SFCJPA is equally focused on non-traditional sources of funding for such projects, including a new regional agency called the San Francisco Bay Restoration Authority, which recently secured $500 million over 20 years to support projects such as SAFER, on the private sector that will rely not just on the direct protection of its assets but also on the protection of roadways to access its assets and electrical and water infrastructure to support it, and finally on opportunities to aggregate flood insurance community wide through private companies and utilize the savings for capital improvements.

Several of these strategies may also bear fruit for the SFCJPA's work along the creek upstream of Highway 101, where it maintains its partnership with the USACE for potential benefits related to federal funding and permitting support. Thus, while local residents are not ready to abandon a federal process that has—after 12 years—not produced tangible results to provide protection almost 20 years after the flood of record, the agency that represents them in this process, the SFCJPA, is exploring all available opportunities to complete this work locally.

5.7 FROM CHANNELS TO CREEKS IN CONTRA COSTA COUNTY

Mitch Avalon

5.7.1 Introduction

This chapter describes the approach that our Flood Control District is taking to ensure long-term, sustainable flood protection for its residents and commercial centers. This approach combines the need for capital replacement of flood protection infrastructure with the natural desire of people to reside in communities with natural creeks rather than traditional flood control channels.

5.7.2 Infrastructure Replacement

Most of our infrastructure was designed and built by the USACE with federal funding during the 1960s and 1970s, a period of rapid development in the County. To receive federal funding, the County had to pay all right-of-way costs, which often resulted in relatively narrow concrete and rip-rap lined channels. There are three reasons why this infrastructure may need to be replaced. First, at some point it will exceed its service life of between 75 and 100 years. Second, the infrastructure has become inadequate to provide the level of flood protection necessary for the community. This could be due to changing land uses within the catchment (e.g., changing from an agricultural land use to an urban land use) or a desired increase in the level of service (e.g., from a 50-year level of protection to 100-year level of protection to meet the US Federal Emergency Management Agency (FEMA) flood insurance requirements). Third, the infrastructure was built with design assumptions that no longer work today. An example would be the past practice of the USACE to design steeper than natural grades in the lower portion of creek channels so they ended up below sea level at the outlet. This was an acceptable design practice at the time because dredging was a standard maintenance operation that was easy and inexpensive. Today, however, flood protection agencies effectively cannot get permits to dredge lower creek channels and the channels fill up with silt, reducing flood flow capacity.

We have an estimated $1.0 billion in infrastructure that we need to replace over the next several decades. Compounding our problem is a severe shortage in funding that barely covers routine maintenance costs. After the passage of a California tax reform measure in 1978, our tax revenue was reduced by 58%. Another measure passed in 1996 requires a vote of all registered voters within the service area, or alternatively all property owners, in order to establish or raise a property tax or assessment. When our infrastructure was originally built by the USACE, federal funding covered up to 90% of the project planning, design, and construction costs. Working with the USACE today to rebuild our infrastructure, only 50% of the overall planning and design costs and between 50% and 65% of the construction costs are covered. In addition to the increase in costs to the local flood control district, there is an order of magnitude higher costs overall in planning, designing, and building a flood control project today compared to a project in the 1960s or 1970s. How will we pay for future projects? We must bring our communities in to help fund infrastructure replacement projects and to provide influence and advocacy at the state and federal levels for increased state and federal funding.

Another consideration in a capital replacement program is life cycle costs. These vary by facility and channel reach. Concrete channels tend to have high initial construction costs, very low ongoing maintenance costs, and high replacement costs. Natural channels require increased right-of-way width (and cost) and generally higher ongoing maintenance but low or zero replacement costs. Natural channels also tend to be more adaptable to changes in the landscape or to climate change impacts than hard facilities. Taking the long view, the costs of natural channels will be less compared to the costs of multiple life cycles for concrete channels.

Our major infrastructure has a remaining service life of 30–50 years. We need to embark now on a planning process for long-range replacement of this essential infrastructure. The question for our communities is this: what type of infrastructure should it be replaced with? Should we simply rebuild our concrete or rip-rap channels, or should they be replaced with more natural systems of vegetation and riparian habitat in a manner that allows natural processes to maintain essential flood protection and water quality improvement functions, recreational and aesthetic values, and flexibility to respond to climate change? Our experience indicates there will be much more public support for replacing existing infrastructure with natural systems. If we pose this question openly, then the answer becomes a community design issue, resulting in community involvement, and ultimately community buy-in and support. This long-range process to engage the community and develop a "creek enhancement plan" was termed the "50-year plan" simply to illustrate the long-range aspect of the process (CCCFCWCD 2009).

5.7.3   Creek Enhancement Planning

The Flood Control District has 116 km (72 mi) of engineered, or historically termed "improved", channels that no longer have the natural features of the original creek. Funding will likely become available to restore some natural features to these channels. In today's world, there is much more funding available, and the award of funding much more likely, for an environmental creek restoration project than for a concrete flood control channel replacement project. As our communities age and land uses change, we will have the opportunity through "redevelopment" of the community to implement more natural flood protection facilities integrated into the new urban landscape (Figs. 5.7 and 5.8).

The Flood Control District can develop Creek Enhancement Plans to, for example, plant riparian vegetation in an earthen channel and still maintain flood protection, if the drainage system is looked at from a catchment perspective, to offset the loss in capacity due to the vegetation planted in

Fig. 5.7 Grayson Creek channel just upstream of Taylor Boulevard, Pleasant Hill, California (Photo by Mitch Avalon 2000)

Fig. 5.8 Remnant of the natural Walnut Creek after Corps project, now referred to as Ellinwood Creek, Pleasant Hill, California (Photo by Mitch Avalon 2000)

the channel. If the goal is to convert a flood control channel to a natural creek, then some Creek Enhancement Plans will need extremely long planning horizons of 50 years or more to achieve all of their objectives. Some plans may be as simple as providing a bypass pipe or an upstream detention basin or increased upstream infiltration to allow a creek section to be natural, while other plans may call for purchasing a row of houses in order to replace a concrete channel with a natural looking creek. These kinds of objectives are achievable and can be implemented without unreasonable disruption to a community if a long-range "50-year" creek enhancement plan is adopted. The Flood Control District will develop these plans if our communities are interested in a more natural environment in our flood protection facilities.

5.7.4   Benefits for the Community

The community gains many tangible benefits to a natural flood protection system in addition to continuing flood risk reduction. Having a natural creek system flow through a neighborhood rather than a concrete channel looks and feels better to the surrounding residents resulting in increased property values. The community can plan and design its public spaces and retail/commercial areas to take advantage of the attraction of a natural system and have a recreational and aesthetic focus along the creek. This can enhance economic activity and investment in the area. A natural creek can provide wildlife linkages between urban ecosystem fragments and open space areas and filter storm runoff to reduce pollutants in the storm-water. As our landscape becomes more urbanized and we have more technological diversions, our children have less opportunity and spend less time interacting in a natural environment. Reestablishing natural creeks in an urban setting will increase opportunities for children to interact with nature, a healthy alternative to the "concrete jungle" (Louv 2008).

The community will have an opportunity for citizens to get involved in creek-related activities, such as clean-ups, water quality monitoring, and fish surveys, or for youth groups to help actively manage portions of the creek by, for example, removing invasive species or trimming vegetation. These activities increase citizen involvement and increase their sense of community. The community can also develop and retain a skilled workforce restoring and maintaining public and private natural creeks. This could include revegetation and soil bio-engineering project work, water quality monitoring, and coordination of erosion prevention/stabilization on private property and stream stewardship training for private property owners. These would be new jobs for the community that can't be out-sourced overseas and help the community's economic sustainability.

5.7.5   Opportunities

There are many opportunities to include long-range planning for the replacement of vital flood protection infrastructure within existing community planning and implementation activities. Each city and county must update its general plan every 20 years. In addition, many cities and counties prepare general plan amendments or specific plans to establish a vision for development of a neighborhood within the city or a county or unincorporated community. Large development projects are required by state law to mitigate their impacts on the surrounding community. Development mitigation measures could include short pieces of channel or creek enhancement with their land-use entitlements. These are all opportunities to include catchment and systemwide creek infrastructure planning into a city's fundamental and foundational planning documents.

There are other drivers for long-range planning that could include flood protection infrastructure on a catchment or systemwide basis. Integrated Regional Water Management Planning in California is a collaboration primarily of water supply, wastewater, and flood protection agencies that could provide funding or cost share contributions to alternative stormwater management approaches that, for example, retain and "harvest" rainfall, thereby enhancing local water supplies for landscape irrigation and reduction of flood peaks. There are also opportunities to develop catchment or creek enhancement plans and/or implement portions of improvements as an offset to or in lieu of stormwater (NPDES, National Pollutant Discharge Elimination System) or regulatory permit requirements. FEMA is continually updating their floodplain maps and adding properties to the flood hazard area, which triggers the need for flood insurance. This is always a driver for increased public awareness and interest in reviewing catchment or creek infrastructure needs within floodplains. Climate change and sea level rise will also be a trigger for long-range creek planning, especially with the expansion of floodplains and exposure of more properties to flood risk.

5.8   THE RIBEIRA DAS JARDAS STREAM: AN URBAN FLOODPLAIN IN LISBON

Graça Saraiva

5.8.1   Evolution of Flood Management Strategies in Portugal

Floodplain management in Portugal has a long tradition, driven by the unbalanced seasonal distribution of precipitation and runoff of its Mediterranean climate. International Iberian river basins, such as Tagus and Douro, as well as smaller basins entirely within Portugal and small, highly urbanized catchments, have been affected by floods, causing extensive economic damages and, in extreme situations, loss of human lives.

Flood control measures and legislation have been implemented since the beginning of the twentieth century, initially with a strong emphasis on structural measures. The "water sector" was responsible for building flood control structures, namely dams and levees/dikes.

In the 1980s, a non-structural approach for floodplain management emerged, mainly coupled with land-use planning regulations, such as floodplain zoning and development constraints, integrated on the spatial planning process at the local (municipal) level. The fact that very severe flash floods occurred in the highly urbanized region around Lisbon in

1967 and 1983, with many casualties (around 400 in 1967), was an alert for the need to consider flood risks in spatial planning and development control. However, the current trends of urban growth and metropolization have increased flood hazards in the most developed areas, such as the Lisbon region and in the Algarve, where tourism has promoted intense urban sprawl.

At the time of Portugal's joining the European Community in 1986, EU environmental policy concerning water resources emphasized mostly solving water quality problems, with a set of directives that intended to achieve water quality standards. With the adoption of the Water Framework Directive (WFD) in 2000 (Directive 2000/60/EC of October 23, 2000) (EU 2000), member states were required to organize themselves under river basin districts governed by river basin commissions, charged with preparing River Basin Management Plans. In Portugal, the WFD was transposed into national legislation in 2005, through the Water Law (Law no. 58/2005 of 29 December) and related legislation.

However, the WFD didn't specifically address the risk of flooding and the effect of future climate change. The European Commission then developed an approach to manage flood risks at the community level, taking into account the uncertainty of climate change and reinforcing flood prevention as well as flood protection and mitigation. The Floods Directive (adopted in 2007) proposed to "establish a framework for the assessment and management of flood risks, aiming at the reduction of the adverse consequences for human health, the environment, cultural heritage and economic activity associated with floods in the Community" (EU 2007).

Under this Directive, floods also must be managed at a river basin scale, with international cooperation for rivers crossing two or more countries. Three distinct steps are required by the Directive: (1) preliminary flood risk assessment, based on an analysis of past floods with significant adverse consequences—these assessments were due in 2011; (2) preparation of flood hazard maps and flood risk maps, at the level of river basin districts, including several scenarios of probability (low, medium, and high), showing the number of inhabitants and type of economic activity potentially affected—these maps were due in 2013; (3) preparation of flood risk management plans to reduce potential adverse consequences of flooding, including prevention, protection, preparedness, flood forecast, and warning systems. These plans were encouraged to include sustainable land use practices and the improvement of water retention—the plans were due by the end of 2015.

A key feature of the Floods Directive was its implementation within the framework of the WFD (again transposed into national law), so that flood risk management plans were developed by the same competent authorities already established to develop and implement the River Basin Management Plans to encourage consistency with WFD goals.

Portugal adopted this Directive in 2010 (Decree-law no. 115/2010 of 22 October). Each of the ten river basin districts of mainland Portugal adopted a flood risk management plan in 2016, following requirements of the Directive, with information accessible to the general public (APA 2016). Thus, overall responsibility for flood management lies with the Ministry of Environment through the Environment Agency, which oversees the River Basin Districts and River Basin Management Plans (The Ribeira das Jardas/Barcarena is included in the Tagus Basin District).

Climate change scenarios for main river basins forecast the concentration of precipitation in winter and the concentration of heavy precipitation events which are likely to increase flood magnitude, frequency, and risks (Santos et al. 2001). This argues for associating flood protection measures with best environmental options, such as natural retention of floodwaters and green infrastructure to reduce runoff. This strategy can strengthen natural flood management, contributing to the protection and restoration of floodplain and coastal ecosystems, which can function to mitigate climate change impacts. Green infrastructures can be broadly described as approaches that work with nature to reduce flooding and restore natural ability to store or slow down flood waters, planned at a basin or catchment scale (EC 2011).

Green infrastructure concepts have been integrated in the spatial planning process in Portugal at the municipal level since 1999, with the objective to develop continuous "green or ecological networks" or "municipal ecological structure". This structure includes rivers and streams, floodplains, steep areas, and other environmentally sensitive areas, aiming to ensure land uses compatible with environmental and hazard protection. These areas are mapped in the municipal plans and should be developed within these constraints, namely, as green infrastructure corridors, or green areas for leisure and recreation. Among areas recognized for good green infrastructure practice are the cities of Porto and Lisbon, and municipalities along the Ribeira das Jardas/Barcarena west of central Lisbon (Silva et al. 2012).

5.8.2 Revitalizing the Ribeira das Jardas (Cacém, Portugal)

Urban streams in the Lisbon region have been very vulnerable to flash floods in recent decades (1967, 1983, 1997, 2008, and 2011), largely because urban growth pressures in the twentieth century allowed dense settlement areas to spill into floodplains. The north bank of the Tagus estuary west of central Lisbon is drained by a series of small streams flowing from north to south, crossing densely developed areas, still maintaining, in some cases, deeply incised valleys within which many stream reaches have retained relatively natural characteristics. This is the case of Ribeira das Jardas stream, whose catchment covers an area of approximately 35 km^2 and originates in the mountainous slopes of vale de Lobos, in the municipality of Sintra, flowing into the municipality of Oeiras and discharging into the Atlantic (Fig. 5.1 b–c). In its 18.8 km, the Ribeira das Jardas flows through the cities of Cacém and Barcarena, before debouching into the Tagus estuary. In its northern upstream reach, it is called the Jardas; below the town of Barcarena, it is known as the Barcarena.

The tremendous potential for restoration of the Ribeira das Jardas for both ecological and social values has been recognized (Saraiva et al. 2001; Silva et al. 2004; Kondolf et al. 2010), and it was here that the first river rehabilitation project in Portugal under the concept of green infrastructure was implemented (Fig. 5.9). A large program of urban renewal, the Polis program, was launched by the Ministry of Environment in 2000, with goals of urban rehabilitation and environmental regeneration in cities and to improve their quality of life. This program was applied in the city of Cacém in 2005 to rehabilitate the Jardas stream and create a greenway along its banks. The Polis program reinforced the coordination between central institutions, notably the Water Agency, in charge of water and river management (now subsumed into the Environment Agency) and local authorities (municipalities), encouraging coordination, shared decision making, and providing special funding (Partidário and Nunes Correia 2004).

Prior to the project, the Jardas was highly altered and degraded, canalized within concrete walls, and consequently, with low habitat complexity and reduced amenity and recreational values (Kondolf et al. 2010). With the selection of the city of Cacém for inclusion in the Polis program, a master plan was developed, establishing the aims of achieving identity, mobility, and sustainability, as well as of creating urban quality of life. The master plan called for a public green area along Jardas stream and the

Fig. 5.9 Ribeira das Jardas rehabilitation project in downtown Cacém, Portugal (Photo by Graça Saraiva 2008)

rehabilitation of the concrete channel in the city center into a green corridor. Importantly, the sewage system treatment was upgraded, and consequently, water quality improved, creating conditions favorable for restoring some ecological functions.

The river rehabilitation project had goals of providing flood protection, regenerating aquatic and riparian habitat, providing space for leisure and recreation, and enhancing scenic and aesthetic values. Continuity along the river corridor was pursued, leading to the demolition of some buildings that had been restricting the floodplain. In addition, measures were implemented to increase permeability and restore riparian vegetation (NPK 2011).

Due to the highly constricted space available, the park was conceived as a set of overlapping systems at different levels, such as riparian vegetation, the circulation network, the green and permeable areas, and the bank structures. Given the lack of space and the extent of prior disturbance, there was no attempt to restore a natural channel. A system of gabions provided flexibility to increase flood conveyance capacity to accommodate the 100-year flood, and to eliminate flow restrictions, and also to create diversity and enhance physical habitats.

Banks and terraced floodplains became permeable land, with riparian vegetation planted to restore a riparian ecosystem. Multiple terraces made possible the creation of detention areas. A network of trails, cycling lanes, and paths has attracted people from the surrounding dense neighborhoods to meet together near the "rediscovered" stream. One of the major benefits of this project was its successful engendering of public use and recreation. Before the project, the area around the Jardas was derelict, the concrete channel hemmed in by fences, hidden between buildings, ignored by the residents, and collecting litter. After the rehabilitation, social uses of the greenway were the most visible benefits. In a survey conducted in 2009 (Kondolf et al. 2010), results showed that 77% of the respondents felt positively about the intervention, considering improvements on river accessibility, pollution, and urban aesthetic quality. Survey responses pointed to the need for more shaded area, as the trees had been only recently planted, as well as more recreation facilities. One ironic result of the survey was that respondents were more critical of the water quality than had been the case prior to the project. In reality, the water quality had measurably improved thanks to the upgraded waste water treatment facilities, but the improved visibility of the stream evidently drew public attention to the water quality (Kondolf et al. 2010).

REFERENCES

APA. 2016. *Planos de Gestão dos Riscos de Inundação*. In: http://www.apambiente.pt/index.php?ref=16&subref=7&sub2ref=9&sub3ref=1250.
Arnold, J. 1988. *The Evolution of the 1936 Flood Control Act*. Fort Belvoir/ Washington, DC: Office of History, US Army Corps of Engineers.
Carter, N.T., and C.V. Stern. 2010. *Army Corps of Engineers Water Resource Projects: Authorization and Appropriations*, Congressional Research Service Report for Congress. R41243.
CCCFCWCD (Contra Costa County Flood Control and Water Conservation District). 2003. *Contra Costa County Watershed Atlas*. Martinez: CCCFCWCD (Contra Costa County Flood Control and Water Conservation District).
CCCFCWCD (Contra Costa County Flood Control and Water Conservation District). 2009. *The 50 Year Plan "From Channels to Creeks"*. Contra Costa County Flood Control and Water Conservation District.
EC (European Commission). 2011. *Towards Better Environment Options for Flood Risk Management*. DG ENV D.1. 236452, European Commission.
EU (European Union). 2000. *Directive 2000/60/EC of the European Parliament and of the Council of 23 October 2000 Establishing a Framework for Community Action in the Field of Water Policy*. Brussels.

————. 2007. *Directive 2007/60/EC of the European Parliament and of the Council of 23 October 2007 on the Assessment and Management of Flood Risks.* Brussels.

Kondolf, G.M., K. Podolak, and A. Gaffney, eds. 2010. *From High Rise to Coast: Revitalizing Ribeira da Barcarena.* Berkeley: Department of Landscape Architecture & Environmental Planning, University of California.

Kondolf, G.M., K. Podolak, and T.E. Grantham. 2013. Restoring Mediterranean-Climate Rivers. *Hydrobiologia* 719: 527–545. https://doi.org/10.1007/s10750-012-1363-y.

Louv, R. 2008. *Last Child in the Woods: Saving Our Children from Nature Disorder.* Chapel Hill: Algonquin Books.

MCFCD (Marin County Flood Control District). 2017a. http://www.marinwatersheds.org/ross_valley.html. Accessed 30 Jan 2017.

————. 2017b. http://www.marinwatersheds.org/flood_stormwater.html. Accessed 30 Jan 2017.

————. 2017c. http://www.marinwatersheds.org/documents_and_reports/documents/CorteMaderaCreekFCSASigned_000.pdf.

————. 2017d. http://www.marinwatersheds.org/documents_and_reports/USACE CorteMaderaCreekProject.html.

————. 2017e. http://www.marinwatersheds.org/documents_and_reports/documents/CorteMaderaCreekProjectupdateflyerNovember2016.pdf.

NPK Consultants. 2011. *Ribeira das Jardas Linear Park in Cacém, Portugal.*

Partidário, R., and F. Nunes Correia. 2004. Polis—The Portuguese Programme on Urban Environment. A Contribution to the Discussion on European Urban Policy. *European Planning Studies* 12 (3): 409–423. https://doi.org/10.1080/0965431042000195001.

Pinto, P.J., and G.M. Kondolf. 2016. Evolution of Two Urbanized Estuaries: Environmental Change, Legal Framework, and Implications for Sea-Level Rise Vulnerability. *Water* 8: 535. doi:https://doi.org/10.3390/w8110535. Online at http://www.mdpi.com/2073-4441/8/11/535/pdf.

RDG (Restoration Design Group). 2013. *Walnut Creek Watershed Inventory.* Prepared for the Walnut Creek Watershed Council. Berkeley: RDG (Restoration Design Group).

Roos-Collins, R. 2007. Bankside San Jose. In *Rivertown: Rethinking Urban Rivers*, ed. P.S. Kibel, 111–141. Cambridge, MA: MIT Press.

Samet, M. 2007. Bankside Federal. In *Rivertown: Rethinking Urban Rivers*, ed. P.S. Kibel. Cambridge, MA: MIT Press.

Santos, F.D., K. Forbes, and R. Moita. 2001. *Climate Change in Portugal. Scenarios, Impacts and Adaptation Measures: Executive Summary and Conclusions.* Lisboa: Fundação Calouste Gulbenkian.

Saraiva, M.G., P. Simões, J. Alves, G.M. Kondolf, and K. Morgado. 2001. *Estudo de Requalificação Paisagística e Ambiental das Ribeiras da Costa do Estoril.* SANEST, Saneamento da Costa do Estoril, SA.

SCVWD (Santa Clara Valley Water District). 2008. *Permanente Creek Flood Protection Project Planning Study Report*, Santa Clara Valley Water District, July.

Silva, J.B., M.G. Saraiva, I. Loupa-Ramos, F. Bernardo, and F. Monteiro. 2004. *Classification of the Aesthetic Value of Jardas Stream*, Deliverable 4-3, URBEM (Urban River Basin Enhancement Methods) Project, EVK-CT-2002-00082 CESUR, IST, Technical University of Lisbon.

Silva, J.B., M.G. Saraiva, I. Loupa-Ramos, and F. Bernardo. 2012. Improving Visual Attractiveness to Enhance City-River Integration—A Methodological Approach to Ongoing Evaluation. *Planning Practice & Research* 28 (2): 163/185. https://doi.org/10.1080/02697459.2012.704734.

Stetson Engineers. 2011. *Capital Improvement Plan Study for Flood Damage Reduction and Creek Management for Flood Control Zone 9/Ross Valley*, Report to Marin County Flood Control and Water Conservation District, May.

U.S. Congress. 2014. *Water Resources Reform and Development Act of 2014*, Public Law 113-121.

USACE (US Army Corps of Engineers). 1963. *Design Memorandum No.1, Walnut Creek Project, Contra Costa County, California, Channel Improvement, General Design and Basis of Design for Reach I (Suisun Bay to Willow Pass Road)*. Sacramento.

———. 1964. *Design Memorandum No.2, Walnut Creek Project, Contra Costa County, California, Channel Improvement, Basis of Design—Reach II, Willow Pass Road to Rudgear Road.*

———. 1972. *Walnut Creek Project, Contra Costa County, California—Letter Supplement No.3 to Design Memorandum No.1*, Prepared by Diversion Engineer, South Pacific.

Williams, P.B. 1990. Rethinking Flood-Control Channel Design. *Civil Engineering* 60 (1): 57–59.

Wong, P.L.R. 2014. *Federal Flood Control Channels in San Francisco Bay Region—A Baseline Study to Inform Management Options for Aging Infrastructure*, PhD Dissertation. University of California Berkeley.

CHAPTER 6

Managing Floods in Urban Catchments: Experiences in Denver Area (Colorado, USA) and Geneva (Switzerland)

Bill De Groot, David Mallory, Georges Descombes,
G. Mathias Kondolf, and Anna Serra-Llobet

Abstract As Denver, Colorado, began its rapid growth in the 1960s, two major floods led to the formation of an extensive flood control district, which adopted a two-pronged approach: Through land-use regulation, it prevented further development in floodplains. In already-built floodplain

B. De Groot
Denver Urban Drainage and Flood Control District, Denver, CO, USA

D. Mallory
Denver Urban Drainage and Flood Control District, Denver, CO, USA

G. Descombes
ADR Architects, Geneva, Switzerland

G. M. Kondolf
University of California Berkeley, Berkeley, CA, USA

University of Lyon, Lyon, France

A. Serra-Llobet (✉)
University of California Berkeley, Berkeley, CA, USA

Aix-Marseille University, Marseille, France

© The Author(s) 2018   135
A. Serra-Llobet et al. (eds.), *Managing Flood Risk*,
https://doi.org/10.1007/978-3-319-71673-2_6

areas, it removed houses where possible and elsewhere used structural measures to reduce flood hazard. The result is fewer houses vulnerable to flooding than before, and large riparian parklands, providing ecological and social benefits. The River Aire flows north and east from the French Alps to central Geneva. About 5 km of the river was channelized in the nineteenth century to provide better drainage, but this conveyed floods faster, increasing flood risk downstream. Restoration of the Aire River included two extensive, shallow impoundments on the floodplain to reduce peak flows downstream and a highly innovative approach to allowing for a dynamic channel within a framework of multiple possible channels from which the river could adopt a course.

Keywords Flood risk management • Urban catchments • *espace de liberté* • Denver • Geneva

6.1   Introduction

G. Mathias Kondolf and Anna Serra-Llobet

Both Denver and Geneva are located in the midcontinent at the foot of massive mountain ranges. The city of Denver's population of 680,000 lies in the heart of the metropolitan area of 2.8 million souls, a sprawling city at the foot of the Rocky Mountains. The canton of Geneva and its immediate suburbs have a population of 1 million, 1.25 million including commuter suburbs. By virtue of their proximity to high mountains, the main rivers are dominated by snowmelt, although Lake Geneva moderates flows in the Rhône, which emanates from it. By virtue of their impermeable surfaces, both cities are prone to urban flooding, from local runoff that is in excess of the limited infiltration capacity of the largely paved-over urban areas (Fig. 6.1).

Denver is mostly a recent creation, built by developers in subdivisions that have spread out farther and farther from the city center, both up the foothills and out onto the infinite plains. As such, in the 1960s and 1970s there existed an opportunity to influence the nature of future urbanization. This opportunity existed elsewhere, but Denver took unusual advantage, thanks to the political opening provided by large floods in 1965 and 1969, coupled with the emergence of environmental planning concepts at the time. In addition to storing flood waters as one of several structural solutions to reduce risk to houses already in vulnerable sites, local governments implemented pro-active regulation and public education to prevent badly

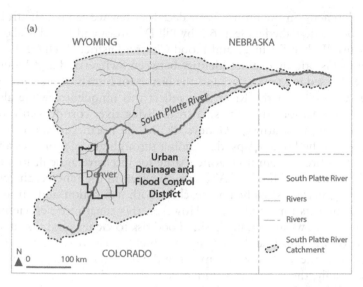

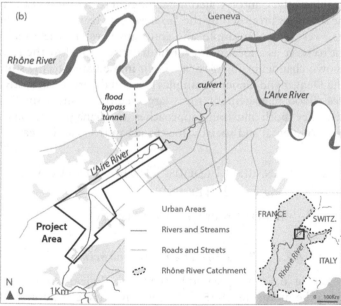

Fig. 6.1   Location maps for the Denver Urban Drainage and Flood Control District in Colorado (USA) within the South Platte River catchment (**a**), and for the Aire River, Geneva (Switzerland) (**b**) (Source: (**a**) adapted from Denver Urban Drainage and Flood Control District, (**b**) adapted from Kondolf 2012, used by permission)

sited developments in the future. Four decades later, we can see the fruits of this effort, as described in Sect. 6.2 by Bill De Groot and David Mallory of the Denver Urban Drainage and Flood Control District (UDFCD).

Geneva's historically distinct identity between Switzerland and France is today expressed in many ways, and the history of human alteration and ongoing restoration of the River Aire reflect this uniqueness while also providing a model for other cities to create living river corridors in their midst. The Aire, in southwest Geneva, drains the northern flank of the Salave range in the French Alps, descending through glacial moraines and thence onto a broad plain now zoned large for intensive agriculture. To improve drainage, much of the Aire was canalized in the nineteenth century, creating perfect straight-line reaches, both a function and an aesthetic reflecting its time and place. However, this canal more efficiently delivered water downstream, increasing flood risk to Geneva and threatening an important industrial district that was built in part over a 1960s-era tunnel carrying the lowermost kilometer of the Aire to its confluence with the Arve. (Sadly, the tunnel was undersized.) In seeking solutions to the flooding problem, the Canton called for ecological restoration of the river as well. The solutions proposed by the *Superpositions* team, led by Georges Descombes, were at once simple and innovative: retain the canal and transform it to serve as a focal point for human use and allow the river to develop its own channel in an *espace de liberté* immediately to the south.

While the Denver and Geneva examples presented here are at vastly different scales (flood management for a large region versus a small river restoration), they both illustrate approaches to managing floods and restoring environmental and social values to urban rivers and streams.

6.2   Managing Floods: The Denver Area Experience

Bill De Groot and David Mallory

6.2.1   *The District and Its Financing*

The UDFCD (Fig. 6.1a) was created by the Colorado legislature in 1969, in response to the disastrous 1965 floods, which attracted national attention. The proposed legislation would probably not have passed were it not for the 1969 flood, which occurred during the legislative session. Today the District serves the city of Denver, the urban portions of six other counties, and 33 incorporated cities and towns. The UDFCD covers 4200 km², with 2500 km of major drainageways. The current population is approximately 2.7 million people. The Board of Directors is composed of locally elected officials (mayors, county commissioners, etc.) appointed by various mechanisms and two registered professional engineers (PEs). This has been key because most board members are elected officials, so they know budgets, conflicting pressures, and so on, but they didn't run for this particular job. They meet ten times per year where the staff has their undivided attention.

UDFCD funding comes from property taxes. Initially, the UDFCD was authorized a millage rate of 0.1 (i.e., a tax rate of $0.0001 per $1 of assessed valuation over all the properties within the district) for operations, then added 0.4 for design and construction, then 0.4 for maintenance, and finally 0.1 for the South Platte River, which adds up to 1.0 mill (i.e., $0.001 per $1 assessed valuation). A constitutional amendment now limits the total mill to 0.6–0.7 mills. Because the UDFCD is a single-purpose district, the funds can't be diverted to other uses.

6.2.2   District Activities over Time

The first major activity of the UDFCD was to inventory drainage basins and sub-basins to determine the extent of problems and to develop a plan to resolve those problems. The initial study indicated that approximately 26% of the major drainageway miles within the District were already developed and needed structural solutions, and the remaining 74% were undeveloped and amenable to preventive approaches. A Master Planning Program was also begun. In probably the most important policy decision in its history, the UDFCD Board made a commitment to develop a comprehensive floodplain management (FPM) program to prevent new problems from being created by new development, while 'fixing' existing problems. They did that by creating the FPM program and the Design and Construction Program at the same time. Over time, this policy became known as the two-pronged approach to FPM.

In the early years, the FPM program emphasized the mapping of 100-year floodplains along undeveloped drainageways in the path of development. Future-conditions hydrology was used to anticipate how urbanization would increase discharges. Other early activities included working with the local governments to adopt floodplain regulations and join the National Flood Insurance Program. In the early years, the Design and Construction Program began implementation of portions of early master plans. These projects were the traditional, structural, single-purpose projects not very friendly to their neighbors or the natural and beneficial functions of the stream corridors. At the same time, on the FPM program side, developers and local governments, while now respecting the 100-year discharge, were also implementing similar single-purpose projects.

We soon realized that single-purpose flood control projects designed for rare events were (by definition) rarely utilized for their intended purpose, so we began to work with our local partners to make flood control projects more useful to their constituents on a daily basis (Fig. 6.2). We informally called this our 'good neighbor policy'. This was in the context of increasing the recognition of concepts of multiple-use or multi-objective management. In 1980 the UDFCD adopted a Maintenance Eligibility Policy: 'Facilities constructed by, or approved for construction by, a local public body after March 1, 1980, must be approved by the (UDFCD) in order for these facilities to be eligible for UDFCD maintenance assistance'. This was an attempt to get a better handle on facilities being constructed by developers and was run by the FPM program. By the mid-1980s we saw an improvement in projects completed by the private sector and by

Fig. 6.2 Stream stabilization project on the South Platte River at Oxford Ave, Englewood, Colorado, September 2016 (Courtesy of the Denver Urban Drainage and Flood Control District)

the Design and Construction Program. We created a 'good examples' page on our website and tried to work with developers and local governments to push these types of projects. We gave preference to floodplain preservation in the Maintenance Eligibility Program.

In many cases, we still weren't getting quality projects from the private sector so we developed an approach to working with developers and the local government permitting authorities. We recognized that developers are in the business to make money, so we thought, 'let's show them how to make money by preserving the floodplain instead of destroying it'. We also recognized that local governments depend on development to provide tax revenues and jobs, which tends to skew their view toward approving development proposals that are sometimes damaging to the floodplain. So 'let's show them how to have development that provides the tax revenue and jobs, but also uses the floodplains to provide amenities to the community'.

6.2.3 Communicating the Vision

We saw the opportunity to prepare a brochure, which would market the floodplain as an asset to developers and communities that could be distributed early in the planning process, for instance, at a pre-application meeting. We chose a trifold brochure with a mini CD. The CD contains five business cases that demonstrate the financial value of preserving floodplains as assets to the development and the community at large. It also contains many other examples of both preservation and restoration projects. The brochure can be accessed at the UDFCD's website at www.udfcd.org under the Downloads tab. It's under the Services/Development Review tab (http://udfcd.org/wp-content/uploads/uploads/resources/floodplain%20management/good_examples_brochure.pdf). The video is under the Resources/Video.

The Information Services and Flood Warning Program hosts internal and external communication, and a robust flood forecasting and warning effort. And, the Floodplain Management Program, in addition to the activities described above, is a Cooperating Technical Partner with the Federal Emergency Management Agency (FEMA), the first such cooperating partner in the country. We review proposals to modify the FEMA floodplain maps submittals within our service area in addition to completing four flood insurance map modernization projects and four flood insurance map maintenance projects.

These concepts were encapsulated in the District's mission statement and vision statement:

Mission Statement: 'The Urban Drainage and Flood Control District works with local governments to address multi-jurisdictional drainage and flood control challenges in order to protect people, property, and the environment'.

Vision Statement: Achieve a sustainable network of safe, efficient, and environmentally sensitive drainage and flood control facilities to best serve an urban community that is aware of its flood risks. Lead the region and the nation by implementing innovative thinking and technology, and by promoting wise use of public and private lands, while providing unsurpassed service to the community.

6.2.4   Conclusions

Since 1969 the population of the UDFCD has tripled, and yet we estimate that there are 5000 fewer structures in the mapped 100-year floodplains. This is due to the two-pronged approach adopted in 1971. The FPM program, working with its local government partners, has been successful in keeping new development out of the floodplain or at least requiring adequate mitigation of the flood hazard. The Design and Construction Program, again working with local government partners, has constructed drainage and flood control facilities with a present worth of $650 million.

Today the UDFCD has a very active Master Planning Program, doing both remedial and preventive planning, with 140 master plans completed. The Design, Construction and Maintenance Program is doing approximately $20 million in drainage and flood control projects and $9 million in maintenance activities per year. The emphasis is on multi-use, promoting the natural and beneficial functions, strengthening the natural systems, and minimizing structural approaches, especially levees. Working together with communication and outreach, these programs have made a significant difference in preserving, protecting, and restoring the natural and beneficial functions of floodplains in the Denver area.

We believe the concept of wise use must include good FPM policy (loss reduction) and promote and protect floodplain resources. We propose the following priorities for wise use of floodplains. (These are the authors' priorities and have not been adopted by the UDFCD.) First, floodplain should be used for safe conveyance of floods and storage of floodwaters. In between floods (which is most of the time), floodplains can be used for natural and beneficial functions such as water quality improvement, groundwater recharge, wildlife habitat, human recreation (parks, trails), agriculture, and other uses compatible with periodic flooding, such as golf courses.

6.3 THE AIRE RIVER: AN *ESPACE DE LIBERTÉ* IN URBAN GENEVA

Georges Descombes

The project addresses a concrete canalized river that was made for flood control in the end of the nineteenth century and then up to the mid-twentieth century. The watershed of the Aire river—approximately 100 km² —is 80% in France, typical of many rivers of the State of Geneva, which shares only 6 km of border with Switzerland's other cantons and 120 km with the French territory (Fig. 6.1b).

The project is part of the 'Master Plan for Geneva 2015' and of the program of revitalization of urban streams. Common goals are shared by Swiss and French administrations concerning quantity and quality of the river water, and the collaboration is organized through a complex set of studies, regulations, and actions summarized in a 'river contract' engaging both countries. The project is at the heart of vast territorial reorganizations at work in the Aire floodplain, transformations aiming at establishing a new balance between urban development, agricultural production, and a restoration of natural ecosystems of the river. In particular, the program aims to reconstruct a web of natural sites along the streams, forming a set of 'linear gardens' up to the very heart of the city. It also recommends giving free public access to the river-banks, but in a way that the pressure on the natural environment is within acceptable levels.

The landscape structure proposed by the project assures protection against floods, the presence of biological corridors, and the capacity to accept the presence of social and productive activities. This true restoration of the territory reconstructs a landscape organization which has today almost entirely disappeared, but is easily readable on historical maps and documents, as well as on a few remains and traces still perceptible on the territory. It was on a patient and accurate reading of the evolution of this territorial context that the project developed its proposals for modifications.

The project was designed for a competition organized in 2001 by the State of Geneva. Instead of destroying the existing canal and simply putting the river in former meanders—as was suggested by the brief and assumed to be correct by most of the other competitors—our team *Superpositions* won the competition with a totally different approach. We displaced the river in a parallel space with the same wavelength of the former meanders and gave the river a new open space (*espace de liberté*) (Fig. 6.3). The plan kept the canal form to make clear and sensible the shifts introduced into the given context, underlying the transformations and giving a way of measuring what was already there and what has been added. The project selects, modifies, transforms, eradicates, or intensifies. And it is done with radical moves, with a kind of brutality, introducing a shock in the existing context, a shock to renew the attention and give a new perception of the site. And not to fall in the pit of a soft naturalistic approach, but once again establishing clearly in the project what was found and how it was transformed.

One of the project's main challenges was to find a proper way to 'build a river'. There was no question, from the very start of the project, of defining a precise and definite new river bed, and we assumed that we didn't

Fig. 6.3 View eastward of the Aire River in May 2016, showing the repurposed nineteenth-century canal on the left and the new river corridor on the right (photo by Fabio Chironi ©, used by permission)

know precisely where the river will move and how it would design its bed. The only question was to decide the nature of the 'launching elements', how to start, how to open the way to the river. In the first phase of the project in 2008, we just got rid of the earth for a depth of about a meter, and we let the river make its own way, its own design.

After five years we could see the result: a new complex and diversified bed carved by the waters, with a succession of pools and riffles. Life had returned; flowers, grasses, butterflies, birds, and fish, they are all there. Nevertheless, we got complaints about the 'too slow' pace the river took to establish a new diversified bed. Fishermen wanted deeper pools, other environmentalists asked for larger rocks or trees trunks to make the river react and move one way or another. Faced with this impatience, we looked for an alternative way of 'building the new river bed' in the next phase. We had already agreed about having just 'launching elements or devices', and in the end, we proposed a 'percolation pattern': a grid of channels excavated, leaving in between diamond or lozenge-shaped remnants of higher ground (approximately 1 m high), opening a great number of possible paths for the new river flows, an 'open' but precise structure.

We started some experiments with models (the first one with a Swiss chocolate bar), and at the end we proposed to bulldoze the earth as in the first phase and then to dig into silt and gravel the 'lozenge' grid pattern. The grid fits the width of former channel alignments of the meanders as a reasonable dimension to give the river space. This stream is torrential, it doesn't want the engineers' sinusoidal symphony nor the landscape architects English style curves. We can just allow the river to shape itself in the grid. This grid is like an extension of the urban grid in a way. To use the grid is to open the system to an infinity of possible variations or adaptations. To build a 'landscaped or engineered' river is to close future evolution. In fact the river, at the end, will impose its own will! Let's observe that the same simplification, not to say mistake, is made as for designing many river beds as for designing new urban developments. A lack of possible variations because of a too fast, impatient solidification of urban or riverine forms. Rivers recall past states, a lesson. There are others. In the Aire floodplain one can easily read the landscape according to John Dixon Hunt recalling the presence of different types of nature. First, clearly present the original given geomorphology as expressed by the powerful mass of the Mount Saleve announcing the nearby Alps. Then the plain itself, a human-constructed agricultural landscape. And then, could our site along the river become a new linear garden? A garden, a place where is represented and questioned the relationships our culture keeps with the given shapes of the world and what and how we transform it, care for it, spoil it, and destroy it. The Aire River revitalization project aims to be a place of interrogation: a public space, a civic place where citizen are faced with the beauty of the world and its fragility and are urged to take part on the finding of new relationships within this world.

The revitalization project is a first step taken as a contribution to change these relationships. The former canal follows the new river space. It is the place and the testimony of the ongoing changes, of the transformations at work. It is the real reason to keep the form of the canal, its straight line besides the new meanders—a linear garden, not a park, nor a linear picnic area, which would miss the point. The accommodation for the public (benches, tables, fountains) are restricted to a few places, and the linear garden is organized in a series of lawns and water gardens, the whole area playing the role of a buffer between this 'rural public space' and the new river meanders.

To renew a river starting in the middle—the entire Aire watershed exceeds our field of operation, upstream and downstream—is not ideal, but one must consider that the project design breathes larger than the 5-km length of the site of intervention. The project has been largely influenced by consideration of its past reality. There is also a considerable improvement over the present situation, with improved protection against floods for the new central district of Geneva downstream. It may also incite changes in the watershed upstream in France.

REFERENCE

Kondolf, G.M. 2012. The *espace de liberté* and Restoration of Fluvial Process: When Can the River Restore Itself and When Must We Intervene? In *River Conservation and Management*, ed. P. Boon and P. Raven, 225–242. Chichester: John Wiley & Sons.

CHAPTER 7

Conclusions

Anna Serra-Llobet and G. Mathias Kondolf

Abstract Flood risk management is critically important to the well-being and economies of our societies, and with increasingly severe weather patterns now manifesting across the globe, flooding issue will gain importance. Experience reflected in the case studies presented in this book demonstrates that the threat of flooding cannot be effectively dealt with by structural methods to reduce hazard alone. An integrated approach that includes reducing vulnerability is key, and integrating multiple benefits in flood risk management projects can increase public support and provide additional funding sources for what are often expensive projects beyond the normal budgets of the responsible public agencies. Every setting is unique, whether the climate be continental or Mediterranean, whether the flood hazard comes principally from large rivers overflowing onto their floodplains or from inadequate urban drainage, and in the envi-

A. Serra-Llobet (✉)
University of California Berkeley, Berkeley, CA, USA

Aix-Marseille University, Marseille, France

G. M. Kondolf
University of California Berkeley, Berkeley, CA, USA

University of Lyon, Lyon, France

© The Author(s) 2018   147
A. Serra-Llobet et al. (eds.), *Managing Flood Risk*,
https://doi.org/10.1007/978-3-319-71673-2_7

ronmental and social resources at stake. As agencies seek to implement integrated flood risk management, they must work within institutional constraints. This has motivated a range of innovative responses, many of which are captured here in the contributions to this book.

Keywords Integrated flood risk management • Structural measures • Non-structural measures • USA • EU

Recent decades have seen increased recognition of the need for a more integrated flood risk management approach that encompasses actions in all parts of the flood risk management cycle (Fig. 1.1), rather than conventional 'flood control', focusing only on structural measures to reduce flood hazard. This book presents experiences of practitioners involved in the 'front lines' of the transition from conventional 'flood control' to flood risk management. The experiences span a wide range of scales and environments in North America and Europe, and provide a window into the challenges and opportunities facing management agencies, planners, NGOs, and the public at large.

Effectively managing floods on large floodplain rivers usually requires coordinated action over spatial scales that commonly transcend administrative boundaries. While the Mississippi presents an example of conflicts between different levels of government within one country, the Rhine is now a model for international cooperation and coordination to reduce flood risk and restore ecosystem values. It is increasingly unlikely that the public and responsible agencies will support single-purpose 'flood control'. Multi-objective projects that increase flood conveyance and enhance ecosystem and social values, such as expansion of flood bypasses along the Sacramento River, are the projects most likely to garner public support and be permitted. Although the scale is smaller, managing urban rivers can pose problems that require innovative approaches to address, such as creation of new administrative arrangements (such as a joint powers authority to overcome the problem of jurisdictions whose boundary is a river). In urban settings, multi-objective projects, such as the restoration of the Ribeira das Jardas near Lisbon, creating a social amenity while improving flood risk management, are most likely to be accepted by the surrounding communities and to receive funding.

One key difference between the conventional engineering approach to 'flood control' and flood risk management is that the former defines a 'design flood' and builds structures to control floods just up to this level, but essentially pretends that greater floods will not occur. In the USA, the

National Flood Insurance Program (NFIP) focuses on the '100-year flood' (i.e., the flood with a return period of 100 years, or a 1% chance of occurring in any given year) and specifies the 100-year "floodplain" as the zone within which homeowners must purchase flood insurance (if their mortgages are federally backed, which is usually the case). By mapping only the extent of this floodplain and considering only this in its determination of who must purchase insurance, the NFIP creates a bimodal condition of being 'in the floodplain' or out of it. As a consequence, most flood control projects in the USA are designed to contain the 100-year flood, not more. Yet even if flood control structures work perfectly to control the 100-year flood, the 'residual risk' of larger floods is still substantial (Plate 2002; Ludy and Kondolf 2012). The concept of 'living with floods' recognizes that complete safety against flooding is an illusion, and this insight creates greater incentive to reduce vulnerability and to provide more 'room for the river' (Merz et al. 2010; ISDR 2004). Under the EU Floods Directive, member states are required to map three different flood levels, and some have opted to map the 'geomorphological floodplain', which shows the natural floodplain that would be flooded if levees and other hydraulic structures fail (Montané et al. 2015; METLTM et MEDD 2004), similar to the 'natural floodplain' proposed by Coulton (2014).

One of the key differences between the North American and European experience pertains to governance. With the implementation of the EU Floods Directive (since 2007), Europe has moved ahead in terms of flood risk management, with the creation of flood risk management plans, which are integrated within the river basin management plans under the Water Framework Directive. While there have been some efforts to implement integrated flood risk management in the USA, such as the Integrated Regional Water Management (IWRM) Program in California (Serra-Llobet et al. 2016), there is no systematic, national-level program to guide individual states or localities to understand their true flood risk and the range of potential measures to reduce the risk. Despite massive investments in flood control structures and implementing the NFIP (and its revisions), the national government in the USA has been unable to prevent further development in flood-prone areas (as recommended by expert panels) and thus unable to stem increases in flood losses (Galloway 1994).

An illustrative governance problem is highlighted in Chap. 2. The Mississippi River and Tributaries (MR&T) Project relies on flood bypasses (locally termed 'floodways') as key components of the project to accommodate flood flows, by inundating designated areas of the floodplain to reduce river stage and prevent flooding of cities elsewhere. In submitting his plan for the MR&T to Congress, General Jadwin observed, 'Man must not try to restrict the Mississippi River too much in extreme floods. The river will break any plan which does this. It must have the room it needs, and to accord with its nature must have the extra room laterally' (U.S. House Doc. 90, 1927; as cited by Shadie and Kleiss 2012). Thus, the MR&T included flood bypasses and backwaters (areas where floodwaters from the main river could back up and pond until the main river stage dropped and the water would drain back out to the main channel). Yet despite the national policy reflected in the operations plan for this complex, multi-state flood control project for the nation's largest river, local interests in Missouri actively resisted using the New Madrid bypass in 2011, taking their petition to the Supreme Court (which denied it). Not so much in the limelight but also undermining the project, local governments in Louisiana have issued building permits for thousands of new houses within the footprint of the West Atchafalaya bypass since 1970, compromising future use of the bypass (Sect. 2.3).

When comparing the European and North American practice now, both are clearly oriented toward meeting the requirements of their respective legislative directives. In the USA, the NFIP dominates flood management actions and influences land-use planning in some predictable and surprising ways. Residential properties mapped as lying within the 100-year floodplain are required to have flood insurance (if the houses have federally backed mortgages, nearly always the case), while those outside (even a few meters horizontally or a few centimeters vertically) are not. Homeowners mapped as being within the 100-year floodplain complain about having to pay their flood insurance premiums (even though the rates have historically been heavily subsidized), and political leaders try to 'get them out of floodplain'—meaning change the map to show fewer houses under water, whether by building levees that reduce the probability of inundation or simply by redoing the hydraulic analysis with additional information to map fewer houses in the floodplain (see Chap. 5 for some examples of how local jurisdictions have dealt with this issue). By contrast, in the EU, a more comprehensive, structured flood risk management approach is now mandated, so that member states all go through the process of assessing flood risk at a catchment scale, mapping hazard and risk for various return intervals, and developing flood risk management plans (EU 2007). Thus, the Floods Directive creates a top-down impetus for each country to develop comprehensive flood risk assessments and measures to reduce risk, rather than simply planning engineering projects to reduce hazard alone. In the USA, even where responsible state and local agencies have adopted an integrated flood risk management approach, they are often held back by the restrictions of the NFIP and lack of mechanisms to coordinate land-use management with flood risk management.

Flood risk management is critically important to the well-being and economies of our societies, and with increasingly severe weather patterns now manifesting across the globe, flooding issue will gain importance. Experience reflected in the case studies presented in this book demonstrates that the threat of flooding cannot be effectively dealt with by structural methods to reduce hazard alone. An integrated approach that includes reducing vulnerability is key, and integrating multiple benefits in flood risk management projects can increase public support and provide additional funding sources for what are often expensive projects beyond the normal budgets of the responsible public agencies. Every setting is unique, whether the climate be continental or Mediterranean, whether the flood hazard comes principally from large rivers overflowing onto their floodplains or from inadequate urban drainage, and in the environmental and social resources at stake. As agencies seek to implement integrated flood risk management, they must work within institutional constraints. This has motivated a range of innovative responses, many of which are captured here in the contributions to this book.

The NFIP was essentially a good idea, and innovative at the time (1968), offering the 'carrot' of subsidized insurance for those already in flood-prone areas balanced by the 'stick' that '...federal benefits [were] contingent on local zoning...' to prevent new development in the floodplain (Houck 1985:78). The NFIP has served as a model and inspiration worldwide over the past 50 years. However, its implementation has been problematic at best. While the experience of the Denver area demonstrates that it was possible for local governments to proactively manage land use on floodplains to reduce flood exposure (Sect. 6.2), in many areas local governments have resisted and undermined restrictions on building in floodplains. As presciently observed in the Unified National Program for Managing Flood Losses of 1966 (White et al. 1966:17): 'A flood insurance program is a tool that should be used expertly or not at all. Correctly applied, it could promote wise use of floodplains. Incorrectly applied, it could exacerbate the whole problem of flood losses.'

The EU Floods Directive is relatively new, so we cannot yet assess the success of its implementation. However, by virtue of its explicit recognition of residual risk and nonstationarity of hydrology, using different scenarios of floods, and encouraging incorporation of socio-economic and climate change in the mapping process, the Floods Directive promotes a more nuanced understanding of flood risk. The flood risk management plans required by the Directive are intended to coordinate stakeholders and agencies at different levels of governance to have a more systemic and holistic management of floods. Its links with the Water Framework Directive encourage integrating ecological values in flood risk management.

Flood damages result from complex interactions among natural and social processes. As the USA considers options to improve the implementation of flood insurance, and as the EU embarks on wide implementation of its new directive, both can benefit by learning lessons from past successes and failures, such as the experiences reported in this volume. The histories of floods and attempts to control them remind us again of the observation that, '[f]loods are an act of God, flood damages result from the acts of men' (White et al. 1966:14). Multi-faceted initiatives recognizing all aspects of the flood risk management cycle, and sharing responsibility at multiple levels of governance (including the individual level), are most likely to result in reductions in flood risk.

REFERENCES

Coulton, K. 2014. Using Soils Data to Map "Natural" Floodplains. *Water Resources IMPACT* 16: 9–12.

EU (European Union). 2007. *Directive 2007/60/EC of the European Parliament and of the Council of 23 October 2007 on the Assessment and Management of Flood Risks.* Brussels.

Galloway, G.E. 1994. *Sharing the Challenge: Floodplain Management into the 21st Century,* Report of the Interagency Floodplain Management Review Committee. Report to Congress. Washington, DC: US Government Printing Office.

Houck, O.A. 1985. Rising Water: The Federal Flood Insurance Program in Louisiana. *Tulane Law Review* 60 (October): 61–164.

ISDR (International Strategy for Disaster Reduction). 2004. *Living with Risk: A Global Review of Disaster Reduction Initiatives.* Geneva: United Nations Publications.

Ludy, J., and G.M. Kondolf. 2012. Flood Risk Perception in Lands 'Protected' by 100-Year Levees. *Natural Hazards* 61 (2): 829–842. https://doi.org/10.1007/s11069-011-0072-6.

Merz, B., J. Hall, M. Disse, and A. Schumann. 2010. Fluvial Flood Risk Management in a Changing World. *Natural Hazards and Earth Systems Sciences* 10: 509–527.

METLTM et MEDD (Ministre de l'équipement, des transports, du logement, du tourisme et de la mer, Ministre de l'écologie et du développement durable). 2004. Circulaire du 21/01/04 relative à la maîtrise de l'urbanisme et adaptation des constructions en zone inondable. BOMEDD n°15 du 15 août 2004.

Montané, A., F. Vinet, T. Buffin-Béranger, O. Vento, and M. Masson. 2015. Cartographie hydrogéomorphologique: émergence d'utilisations règlementaires en France. *Géographie Physicque et Environnement* 9: 37–60.

Plate, E.J. 2002. Flood Risk and Flood Management. *Journal of Hydrology* 267: 2–11.

Serra-Llobet, A., E. Conrad, and K. Schaefer. 2016. Governing for Integrated Water and Flood Risk Management: Comparing Top-Down and Bottom-Up Approaches in Spain and California. *Water* 8: 445. https://doi.org/10.3390/w8100445.

Shadie, C.E., and B.A. Kleiss. 2012. *The 2011 Mississippi River Flood and How the Mississippi River & Tributaries Project System Provides Room for the River.* 2012 EWRI-ASCE World Environmental and Water Resources Congress, Albuquerque.

White, G.F., J.E. Goddard, J.R. Hadd, I. Hand, R.A. Hertzler, J.V. Krutilla, W.B. Langbein, M.J. Schussheim, and H.A. Steele. 1966. A Unified National Program for Managing Flood Losses. Report of the Task Force on Federal Flood Control Policy, submitted by President L.B. Johnson to J.W. McCormack, Speaker of the US House of Representatives, 10 August 1966.

INDEX

© The Author(s) 2018
A. Serra-Llobet et al. (eds.), *Managing Flood Risk*,
https://doi.org/10.1007/978-3-319-71673-2

Shadie, C., 13
Sharing the Challenge: Floodplain Management into the 21st Century, 35
Socio-economic assets, vulnerability of, 6
Special Flood Hazard Areas (SFHAs), 55, 57, 61
SPFC, *see* State Plan of Flood Control
Standard Levee Design, 48–49
Standard levee section, 48
Standard Project Flood, 101
protection, 49, 50, 53
State of California, 46, 67, 120
State of Geneva, 143, 144
State Plan of Flood Control (SPFC), 46–52, 54–56
Steelhead trout, 112
Stormwater management approaches, 126
Strategy to Advance Flood protection, Ecosystems, and Recreation Along the Bay (SAFER Bay), 115, 119–120
Stream stabilization project, 140
Strole, T., 14
Structural measures, flood hazard, 148
Structured flood risk management approach, 150
Superpositions team, 138, 144
Sutter Bypass, 48, 64

T
Tagus Basin, 96
Tagus estuary, 97, 129
Traditional engineering approaches, v, vi
Traité de Versailles, 81
Tributary basin improvements, 15, 19
Tulla, J.G., 77, 80
Twenty-First Century, strategic plan for, 111

U
UDFCD, *see* Urban Drainage and Flood Control District
United States (US), 151
flood control projects in, 149
integrated flood risk management, 149
National Flood Insurance Program (NFIP), 148, 150
national government in, 149
Upper Elkhorn project, 67, 68
Upper Guadalupe River project, 113
Upper Mississippi River, 34
1993 flood, 24
navigation system, 26–27
system, 27, 32, 36
Upper Rhine, 76, 77, 79, 82, 83
fluvial hydrosystem, 80
hydrosystem, 82
restoration of, 83
restoring flood capacity and ecological function along, 79
valley, 4
Urban catchments in San Francisco Area and Lisbon Estuary Area, 95–98
Urban development, 52, 53, 65, 96, 100, 143, 145
in Sacramento Valley, 49
Urban Drainage and Flood Control District (UDFCD), 138–142
Urban expansion in Lisbon, 95
Urban floodplain in Lisbon, 126–131
Urban Geneva, *Espace de Liberté* in, 143–146
Urban levees, 54, 56
improvement projects, 51, 59, 60
systems, 56, 59, 60
Urban level of flood protection, 53, 54
Urban streams, 143
flood protection projects in, 111
in Lisbon region, 7, 129

Printed in the United States
By Bookmasters

Printed in the United States
By Bookmasters